SCUOLA NORMALE SUPERIORE

QUADERNI

Ivar Ekeland

Exterior Differential Calculus and Applications to Economic Theory

PISA 1998

ISBN: 978-88-7642-251-5

Notes by Paolo Guiotto and Tiziano Vargiolu
Scuola Normale Superiore of Pisa

CONTENTS

Foreword

During the academic year 1995-96, I was invited by the Scuola Normale Superiore to give a series of lectures on my work in progress with Pierre-André Chiappori. MM. Paolo Guiotto and Tiziano Vargiolu attended these lectures and reworked the material to write a set of notes, which I then reviewed and expanded.

These lecture notes are thus the result of a joint effort by MM. Paolo Guiotto and Tiziano Vargiolu and myself. The material is taken from the original paper [7] (see also [6]), and the exposition of exterior differential calculus owes much to the beautiful book [4] by Bryant, Chern, Gardner, Goldschmidt and Griffiths (see also [14]). In the past year, there has been further progress on the subject matter, as recorded in [8], [9], and [10].

The purpose of these notes is to make the underlying economic problems and the mathematical theory of exterior differential systems accessible to a larger number of people. Indeed, the articles are mainly concerned with economic theory, and proceed as quickly as possible to the main results, giving only the barest outline of the underlying mathematical theory, while the book [4] is concerned with exterior differential calculus as a whole, and goes much beyond our immediate needs. It is the purpose of these notes to go over these results at a more leisurely pace, keeping in mind that mathematicians are not familiar with economic theory and that very few people, if any, have read Elie Cartan.

2

It is not often that one has the privilege of exploring a subject where deep economics and deep mathematics are so intimately related. I thank the Scuola Normale Superiore for giving me the opportunity to lecture on this work, MM. Paolo Guiotto and Tiziano Vargiolu for the very stimulating discussions we had, and their patience in writing these notes, and my very good friend professor Antonio Ambrosetti for his constant help and encouragement. He had the idea of these notes, and we are glad that he did.

Ivar Ekeland

CHAPTER 1

From economics to mathematics

Many questions in economic theory boil down to one of the following mathematical questions, which we write in order of increasing difficulty:

FIRST QUESTION. Given a smooth vector field X in $\mathbb{R}^n$, and a positive number $k < n$, is it possible to find real functions λ_i and f_i, $1 \leq i \leq k$, such that:

$$(1) \qquad X(x) = \sum_{i=1}^{k} \lambda_i(x) \nabla f_i(x)$$

where ∇f_i denotes the gradient of f_i. Note that the question is trivial for $k \geq n$.

SECOND QUESTION. Can we find the λ_i, f_i in the decomposition (1) such that the λ_i are non-negative and the f_i are convex?

THIRD QUESTION. Can we ask the λ_i, and the f_i in the decomposition to satisfy some additional equations of the following type

$$\phi_j(\lambda(x), \nabla f(x), x) = 0, \quad 1 \leq j \leq p,$$

where the ϕ_j are prescribed functions.

In the very particular case when $k = 1$ and $\lambda \equiv 1$, the answer to the first question is given by the well-known:

THEOREM 1.1. *Given a smooth vector field X in $\mathbb{R}^n$, there exists f such that $X = \nabla f$ if and only if:*

$$\frac{\partial X_i}{\partial x_j} = \frac{\partial X_j}{\partial x_i} \quad \forall i, j = 1, \ldots, n$$

In the general case when λ is a non-constant fuction, the answer to the first question is given by the Frobenius theorem for $k = 1$, and by the Pfaff theorem for $k > 1$. All solutions are local, that is, they are defined and solve (1) on some neighbourhood of a prescribed point $\bar{x}$.

To solve the problem posed in the second and third question, we will need the tools developed at the turn of the century by a score of mathematicians from France and Germany, the most prominent of which is Elie Cartan. In a series of articles and books extending over fifty years, he systematized these tools and discovered a beautiful mathematical structure, which he called exterior differential calculus, where antisymmetry plays a fundamental role. He has given an overview of the theory in a book [5], which is still in print. This theory has been very useful in understanding the local geometry of submanifolds and the differential structure of Lie groups, but has had few applications to other areas of mathematics. It is no exaggeration to claim that it has been little known to analysts, let alone economists. And yet, I think that the Cartan-Kähler theorem, for instance, is as basic a result for partial differential equations as the Cauchy-Lipschitz theorem is for ordinary differential equations.

But before we solve these problems, we have to tell the reader where they come from, and this will require going into basic economic theory. Once the foundation is laid, and some understanding of the basic economic issue is gained, we can proceed with the mathematical analysis. Thus, our hope is to make mathematics and economics walk hand in hand, as so often is the case with physics.

CHAPTER 2

The single consumer model

Let us describe the standard framework of microeconomics (see any textbook on the subject, [15] for instance). The basic behavioural assumption is that *each individual makes the best choice among all opportunities which are open to him.* Mathematically, this translates into the following kind of problem:

$$(2) \qquad\qquad \max_{x \in K} U(x).$$

K is called the *feasible set*. Each $x \in K$ represents a possible choice (or decision) for the individual. Preferences are given by the real-valued function U, the *utility function*. If $x, y \in K$, and $U(x) > U(y)$, then we say that the individual *prefers* x to y; on the other hand if $U(x) = U(y)$, then we say that x and y are *indifferent*. For a given x, the set of all $y \in K$ such that $U(x) = U(y)$ is called the *indifference set* through x.

In economics, decisions or choices concern goods to be consumed (or produced; there is a well-developed theory of production, but we will not go into it in these notes). Then K is a subset of $\mathbb{R}^n$, where n is the number of goods available in the economy. For each $x \in K$, the i-th coordinate x^i represents a quantity x^i of the i-th good; it is assumed that goods are divisible, so that x^i can be any real number, and it is usually assumed also that x^i cannot be negative.

With each good i is associated a price p_i per unit of good, so that a bundle $x = (x^1, \ldots, x^n)$ costs $p_1 x^1 + \ldots + p_n x^n$ – in short, $p \cdot x$. The consumer starts out with a wealth which is normalized to 1, and chooses the bundle of goods he (she) prefers among those he (she) can afford. In mathematical terms, this is stated as a classical problem in optimization:

$$K = \left\{ x \in \mathbb{R}^n : x^i \geq 0, \ \forall i = 1, \ldots, n, \ \sum_{i=1}^{n} p_i x^i \leq 1 \right\}.$$

So the problem (2) becomes

$$(\mathcal{P}_{p,U}) \begin{cases} \max U(x), \\ x_i \geq 0, \\ p \cdot x \leq 1. \end{cases}$$

To make economic sense of this problem, we need some additional hypotheses on U. For instance, we would like the optimal solution to be unique, which leads us to the idea that U should be strictly concave (there are other, perhaps more compelling, economic reasons for assuming concavity). On the other hand, it is clear that the relevant economic object is the preference relation, not the utility function: if h is any increasing function on the real line, U and $V = h \circ U$ define the same preference relation ($U(x) \geq U(y)$ if and only if $V(x) \geq V(y)$), and yet V need not be concave when U is. So we have to enlarge somewhat the notion of concavity:

DEFINITION 2.1. We say that U is (strictly) *quasi-concave* if the sets $\{x : U(x) \geq a\}$ are (strictly) convex $\forall a \in \mathbb{R}$.

REMARK 2.1. Is easy to see that in general a quasi-concave function is not concave. More generally if U is quasi-concave, so is $h \circ U$ for every increasing function $h : \mathbb{R} \to \mathbb{R}$. □
We shall henceforth assume that:

HYPOTHESES 0 (H0).
(i) U is $C^\infty(\mathbb{R}^n)$ and strictly quasi-concave;
(ii) if $\bar{x}$ solves problem (2), then $\bar{x}_i > 0 \ \forall i$;
(iii) $\frac{\partial U}{\partial x_i} > 0, \ \forall i$.

If (H0) holds, we have a unique optimal solution $\bar{x}$. It is characterized by:

$$U'(\bar{x}) = \lambda p, \quad \text{for some } \lambda \in \mathbb{R}^+, \text{ and } \quad p \cdot \bar{x} = 1.$$

This characterization is a standard result in convex analysis, or in optimization theory. The number λ is called the *Lagrange multiplier*.

Note that in general both λ and $\bar{x}$ depend on p. Thus in this situation, we have an optimization problem depending on the parameters $p = (p_1, \ldots, p_n)$ and a map sending p into an unique optimum point $\bar{x}(p)$. The function $p \mapsto \bar{x}(p)$ is called *demand function* of the individual.

It is well-known that if a function U is concave, then $U''(x)$ is negative semi-definite, and that if $U''(x)$ is everywhere negative definite, then U is strictly concave. There are analogous conditions for the quasi-concave case. To state them, we need to introduce a hyperplane $E(x)$, which is in fact the tangent hyperplane to the indifference set hypersurface through x:

$$E(x) := \{y : U'(x) \cdot y = 0\}.$$

Consider the restriction $U''(x)_{|E(x)}$ of the quadratic form $U''(x)$ to $E(x)$. If U is quasi-concave, then $U''(x)_{|E(x)}$ is negative semi-definite. Conversely, if $U''(x)_{|E(x)}$ is negative definite, then U is strictly quasi-concave.

In order to obtain regularity for the demand function we make a further (generic) assumption.

HYPOTHESIS 1 (H1). $U''(x)_{|E(x)}$ is negative semi-definite for all x such that $x_i > 0$ for all i.

Applying the implicit function theorem we obtain the following.

PROPOSITION 2.1. *If* (H0) *and* (H1) *hold, then the demand function is* C^∞.

Again the proof is left to the reader. We are now ready to set the following problem:

Given a C^∞ function $x(p)$, can it be a demand function?

In other words, is it possible to find a quasi-concave function U such that x is the optimal solution to the problem $(\mathcal{P}_{p,U})$?

It may be the case (and in fact, it is) that the answer is no, unless $x(p)$ satisfies additional conditions (in fact, certain systems of nonlinear partial differential equations). Testing these conditions against the data, that is, checking whether the actual demand functions, as observed by econometricians, satisfy these additional conditions, then provides us with a test for the model we started from. In other words, the idea that people maximize their utility over a feasible set, far from flying high up in the air as many people believe, can actually be tested against hard data, just as physicists test their models against experimental data.

So we have to infer the (unobserved) utility function U from the (observed) demand function $x(p)$. There are two ways to do this.

The first one is the so called *direct approach*. We assume that the function $x(p)$ is invertible (which is the case in general). Then, recalling that

$$U'(x(p)) = \lambda(p)p, \quad \lambda > 0$$

we have

$$p(x) = \frac{1}{\lambda(x)} \nabla U(x).$$

Since the left-hand side $p(x)$ is known, and the decomposition on the right-hand side is not, we recognize the problem we stated in the first question of the preceding section (with $k = 1$). This approach has been developed by the Italian mathematician Antonelli [1]. However, his method has fallen into disuse, mainly because $x(p)$ is a more natural function than $p(x)$, even though they are mathematically equivalent.

The second method to solve the problem is called the *indirect approach*. We start by defining the so called *indirect utility function* or *value function*:

$$V(p) := \max\{U(x) : x \geq 0, p \cdot x = 1\}.$$

By this definition we have that

$$V(p) = U(x(p)) = U(x(p)) + \lambda(p)(1 - p \cdot x(p)),$$

due to the fact that $p \cdot x(p) = 1$. Differentiating with respect to p:

$$V'(p) = (U'(x(p)) - \lambda(p)p) \cdot x'(p) + \lambda'(p)(1 - p \cdot x(p)) - \lambda(p)x(p)$$
$$= -\lambda(p)x(p).$$

So we can write

$$x(p) = -\frac{1}{\lambda(p)} \nabla V(p).$$

Since the left-hand side is known, and we are looking for the decompostion on the right-hand side, we are back to the first question again. If we can solve it, and find λ and V, the following proposition will give us U:

PROPOSITION 2.2. *Suppose that* (H0) *and* (H1) *holds. Then we have:*

a) $U(x) = \min\{V(p) : p \cdot x \geq 1\}$,

b) *U is quasi-concave if and only if V is quasi-convex.*

This approach was first developed in Pisa in 1886, by a young researcher in mathematics named Antonelli, who later left academia and went on to a long and distinguished career in engineering. Among other things, in 1923 he promoted Italy's first international airline, the Genoa-Barcelona. His paper [1] (in fact, a booklet published at his own expense) went unnoticed, and it was rediscovered in 1943, one year before the author's death, by Herman Wold, who came across it accidentally while browsing through the personal library of Ladislaus von Bortkiewicz, which had been acquired by the university of Uppsala. In the meantime, of course, the result had been rediscovered again, by another Italian named Slutsky (see [22]). The basic characterization is as follows.

THEOREM 2.1. *Let the map $x \in C^\infty$ satisfy the condition*

$$p \cdot x(p) = 1 \quad \forall p .$$

Then $x(p)$ is a demand function if and only if the matrix $\Sigma = (\sigma_{ij})_{ij}$, defined by:

$$\sigma_{ij} := \frac{\partial x_i}{\partial p_j} - \sum_k p_k \frac{\partial x_i}{\partial p_k} x_j$$

is symmetric and positive definite.

There are several proofs of this result (see for instance the textbook by Kreps [15] for the most popular one). We will prove it by using the Frobenius theorem, which will give us a first glimpse into exterior differential calculus.

CHAPTER 3

Exterior differential calculus

It is now time to point out that the three questions we asked in section 1 are not correctly formulated. Indeed, formula (1) is not invariant by a change of coordinates. If we set $y = \Phi(x)$, with Φ a diffeomorphism, the vector field $X(x)$ becomes $T_x\Phi(x)X(x)$, while $T_x\Phi(x)\nabla f(x)$ is *not* the gradient of $\nabla f(\Phi^{-1}(y))$ at y. So the two sides of formula (1) behave according to different rules, and equality can hold only in one particular set of coordinates.

This is because the gradient is not the right notion. It requires one to identify a linear form with a vector via an Euclidean structure, which makes changes of coordinates very complicated. We shall forget this identification, and work directly with linear forms, without identifying them with vectors; more generally, we will work directly with differential forms (fields of linear forms) without identifying them with vector fields. Let us first recall some basic facts, referring to the book by V. Arnold [2] for more.

3.1. – Preliminaries

Given $x \in \mathbb{R}^n$ let us denote by $T_x\mathbb{R}^n$ the *tangent space* to $\mathbb{R}^n$ at the point x. Each element of $T_x\mathbb{R}^n$ is called a *tangent vector*, and we will set, as usual,

$$TR^n := \{(x, v) : x \in \mathbb{R}^n, \ v \in T_x\mathbb{R}^n\}.$$

A smooth section of $T\mathbb{R}^n$ is called a *vector field*. In other words a vector field X is a mapping defined on $\mathbb{R}^n$ such that $X(x) \in T_x\mathbb{R}^n$. For properties of vector fields we refer the reader to any textbook on differential geometry, for instance the previously cited book by Arnold [2].

Let us denote with $T_x^*\mathbb{R}^n$ the cotangent space at x, that is, the dual of $T_x\mathbb{R}^n$. In other words, an element in $T_x^*\mathbb{R}^n$ is a tangent linear form at x, just as an element in $T_x\mathbb{R}^n$ is a tangent vector at x. Denote by $T^*\mathbb{R}^n$ the cotangent bundle:

$$T^*\mathbb{R}^n = \{(x, \xi) : x \in \mathbb{R}^n, \ \xi \in T_x^*\mathbb{R}^n\}.$$

The canonical coordinate system gives us a natural base for the tangent space $T_x\mathbb{R}^n$: we will denote it by

$$\frac{\partial}{\partial x^1}, \dots, \frac{\partial}{\partial x^n}.$$

This notation follows from our understanding a vector field as a differential operator on functions f defined on $\mathbb{R}^n$:

$$\xi(f) = \sum_i \frac{\partial f}{\partial x^i} \xi^i.$$

Finally we will denote by $dx^1, \dots, dx^n$ the base dual to the previous one in the space $T_x^*\mathbb{R}^n$.

DEFINITION 3.1. A differential 1-form is a smooth section of $T^*\mathbb{R}^n$ over $\mathbb{R}^n$.

In other words, a differential 1-form is a smooth mapping ω defined on $\mathbb{R}^n$ such that $\omega(x) \in T_x^*\mathbb{R}^n$ for all $x \in \mathbb{R}^n$. In canonical coordinates a 1-form ω will be written as $\omega = \sum \omega_i dx^i$, so that if $\xi = \sum \xi^j \frac{\partial}{\partial x_j}$ is a vector field, the operation of ω on ξ is given by

$$\omega(\xi) = \sum \omega_i \xi^i.$$

Now we introduce the more general concept of differential k-forms, and we introduce two fundamental operations: the *exterior product* and the *exterior derivative*.

3.2. – Exterior product

Exterior k-forms can be constructed from 1-forms by a simple algebraic operation, called the *exterior product* :

DEFINITION 3.2. Let $\omega_1, \ldots, \omega_k$ be 1-forms. The exterior product $\omega_1 \wedge \ldots \wedge \omega_k$ is the k-form defined by

$$\omega_1 \wedge \ldots \wedge \omega_k (\pi^1, \ldots, \pi^k) = \sum_\sigma (-1)^{\text{sign}(\sigma)} \omega_1 \left(\pi^{\sigma(1)}\right) \ldots \omega_k \left(\pi^{\sigma(k)}\right)$$

where the sum ranges over all permutations σ of $\{1, \ldots, k\}$.

For instance, for $k = 2$

$$\alpha \wedge \beta (\pi, \chi) = \alpha(\pi)\beta(\chi) - \alpha(\chi)\beta(\pi)$$

$$= \left(\sum_i \alpha^i \pi_i\right) \left(\sum_j \beta^j \chi_j\right) - \left(\sum_j \alpha^j \chi_j\right) \left(\sum_i \beta^i \pi_i\right)$$

$$= \sum_{i<j} \left(\alpha^i \beta^j - \alpha^j \beta^i\right) \left(\pi_i \chi_j - \pi_j \chi_i\right) .$$

Obviously, this is the simplest 2-form related to α and β and satisfying two basic properties, namely bilinearity and antisymmetry.

If M is a n-dimensional manifold, and $(p_1, \ldots, p_n)$ is a local coordinate system on M, then $(dp_1, \ldots, dp_n)$ is a basis for the space of differential 1-forms, and any differential k-form α on M can be written as:

$$\alpha(p) = \sum \alpha_{i_1, \ldots, i_k}(p) \, dp_{i_1} \wedge \ldots \wedge dp_{i_k}$$

where $i_1 < \ldots < i_k$ ranges over all k-subsets of $\{1, \ldots, n\}$. It follows that the exterior products immediatly extends to k-forms:

DEFINITION 3.3. Let α be a k-form, and β be a ℓ-form, then $\alpha \wedge \beta$ is a $(k + \ell)$-form such that

$$\alpha \wedge \beta(p^1, \ldots, p^{k+\ell})$$

$$= \sum_\sigma \frac{1}{k!\ell!} (-1)^{\text{sign}(\sigma)} \alpha \left(p^{\sigma(1)}, \ldots, p^{\sigma(k)}\right) \beta \left(p^{\sigma(k+1)}, \ldots, p^{\sigma(k+\ell)}\right)$$

where the sum ranges over all permutations σ of $\{1, \ldots, k + \ell\}$.

Finally, a few properties of the exterior product must be kept in mind:

- The exterior product is associative
- If α is a p-form and β is a q-form then:

(3) $$\alpha \wedge \beta = (-1)^{pq} \beta \wedge \alpha.$$

- Whenever ω is linear (or of odd order), $\omega \wedge \omega = 0$.
- More generally, if $\omega_1, \ldots, \omega_s$ are linearly dependent 1-forms, then

$$\omega_1 \wedge \ldots \wedge \omega_s = 0.$$

- But: if ω is a 2-form (or a form of even order), $\omega \wedge \omega \neq 0$ in general.
- Finally, for any k-form, $(\omega)^s = \omega \wedge \omega \wedge \ldots \wedge \omega$ is a (ks)-form. In particular, $(\omega)^s = 0$ as soon as $ks > n$.

3.3. – Exterior derivative

We now define the exterior derivative of a differential form. It is first defined for 0-forms (that is, smooth functions) and the definition extends easily to k-forms.

DEFINITION 3.4. Let $U : \mathbb{R}^n \to \mathbb{R}$ a smooth function. Then dU is the 1-form defined by:

$$dU = \sum_i \frac{\partial U}{\partial x_i} dx^i.$$

More generally if

$$\alpha = \sum \alpha_{i_1,\ldots,i_k} dx^{i_1} \wedge \ldots \wedge dx^{i_k},$$

is a k-form, then $d\alpha$ is the $k+1$-form, defined by:

$$d\alpha = \sum d\alpha_{i_1,\ldots,i_k} \wedge dx^{i_1} \wedge \ldots \wedge dx^{i_k}.$$

Note that the coefficients $\alpha_{i_1,\ldots,i_k}$ of the k-form α are just smooth functions, so that $d\alpha_{i_1\ldots i_k}$ is a 1-form given by the first rule, and the definition is complete.

The exterior derivative satisifes the following properties:

(i) d is a linear operator,

(ii) $d^2 = 0$, that is $d(d\alpha) = 0$ for every α,

(iii) $d(\alpha \wedge \beta) = d\alpha \wedge \beta + (-1)^p \alpha \wedge d\beta$, for any p-form α.

Property (ii) is easily checked for $\alpha = dU$ (use formula 3.4), where it amounts to the equality of cross-derivatives: $\frac{\partial^2 U}{\partial x^i \partial x^j} = \frac{\partial^2 U}{\partial x^j \partial x^i}$. It is then extended to all k-forms by formula 3.4.

An important feature of property (ii) is that it has a converse, known as *Poincaré's lemma*: if a k-form α satisifies $d\alpha = 0$ is some neighbourhood of some point x, then there is a (possibly smaller) neighbourhood $\mathcal{U}$ of x and a $(k-1)$-form ω such that $\alpha = d\omega$ on $\mathcal{U}$.

An important feature of the exterior product and the exterior derivative is that they are *natural* operations. This means in particular that they do not depend on the particular coordinate system where the computation are made.

As an example of properties that depend on the coordinate system, let us think of the Hessian of a smooth function U at a point x: it is the symmetric matrix $U''(x)$, with coefficients

$$\frac{\partial^2 U}{\partial x_i \partial x_j} \, .$$

If we perform a change of variables $x = \phi(y)$, the Hessian becomes

$$\sum_{i,j} \frac{\partial \phi_i}{\partial y_k} \frac{\partial^2 U}{\partial x_i \partial x_j} \frac{\partial \phi_j}{\partial y_l} + \sum_i \frac{\partial U}{\partial x_i} \frac{\partial^2 \phi_i}{\partial y_k \partial y_l} \, .$$

If x is a critical point, that is $U'(x) = 0$, then the second term vanishes and the first term tells us that $U''(x)$ behaves like a quadratic form; certain things, like the number of positive eigenvalues, can be attached to the Hessian independently of the coordinate system. In the general case, however, the second term does not vanish, and the only invariant thing that can be said about the Hessian is that it is symmetric. For instance, it may be the case that the Hessian is positive definite in some coordinate system and negative definite in another.

Note that the symmetry of the Hessian is equivalent to the relation $d^2 U = 0$, which states that the antisymmetric part is zero. One can think of exterior differential calculus as an antisymmetric calculus, where symmetric parts are systematically discarded and we are left with the antisymmetric ones, which are the only ones to be invariant under changes of coordinates.

To define naturality of operations in a precise way, we need the concept of *pullback*. It is defined as follows. Let $\phi : \mathbb{R}^n \to \mathbb{R}^m$ a differentiable mapping. The mapping ϕ induces a mapping from differential forms on $\mathbb{R}^m$ to differential forms on $\mathbb{R}^n$: if α is a k-form on $\mathbb{R}^m$ then $\phi^*(\alpha)$ is a k-form on $\mathbb{R}^n$, the action of which is defined by transposition as follows.

If $v_1, \ldots, v_k$ are vectors of the tangent space $T_x\mathbb{R}^n$, then we can construct paths starting at the point x with velocity vectors given just by $v_1, \ldots, v_k$. Then we transfer the paths on $\mathbb{R}^m$ by ϕ and we compute their velocities: let us call $w_1, \ldots, w_k$ these velocities. We set:

$$w_i = T_x\phi v_i \,,$$

$$\phi^*(\alpha)(v_1, \ldots, v_k) := \alpha(T_x\phi v_1, \ldots, T_x\phi v_k).$$

To say that the operations $\wedge$ and d are natural means that the following properties hold:

(I) $\phi^*(\alpha \wedge \beta) = \phi^*(\alpha) \wedge \phi^*(\beta)$;
(II) $d[\phi^*(\alpha)] = \phi^*(d\alpha)$.

3.4. – Differential ideals

Having defined a product between differential forms we can introduce the algebraic concept of *ideal*.

DEFINITION 3.5. A set I of differential forms is called an algebraic ideal if the following conditions holds:

(i) I is a linear subspace;
(ii) if $\alpha \in I$ and β are differential forms, then $\alpha \wedge \beta \in I$.

Moreover we say that the ideal I is generated by $\alpha_1, \ldots, \alpha_p$ (and we will write $I = [\alpha_1, \ldots, \alpha_p]$) if every $\alpha \in I$ is of the type

$$\alpha = \sum_i \alpha_i \wedge \beta_i,$$

for some differential forms β_i.

The following proposition gives conditions for 1-forms to be generators of an ideal.

PROPOSITION 3.1. *Let* $\alpha_1, \ldots, \alpha_p$ *be linearly independent* 1-*forms, and* I *the ideal they generate. Finally let* ω *be a* k-*form. Then the following conditions are equivalent:*

(i) $\omega \in [\alpha_1, \ldots, \alpha_p]$;

(ii) $\omega \wedge \alpha_1 \wedge \ldots \wedge \alpha_p = 0$;

(iii) *if* $\xi_1, \ldots, \xi_k$ *are vector fields on* $\mathbb{R}^n$ *such that*

$$\alpha_i(\xi_j) = 0, \quad \forall i = 1, \ldots, p, \ j = 1, \ldots, k,$$

then

$$\omega(\xi_1, \ldots, \xi_k) = 0.$$

PROOF. (i) $\Rightarrow$ (ii) If $\omega \in I$ then $\omega = \sum_i \alpha_i \wedge \beta_i$, hence, by (3) and the Definition 3.2,

$$\omega \wedge \alpha_1 \wedge \ldots \wedge \alpha_p = \sum_i \beta_i \wedge \alpha_i \wedge \alpha_1 \wedge \ldots \wedge \alpha_p = 0.$$

(ii) $\Rightarrow$ (i) Let us complete $\{\alpha_1, \ldots, \alpha_p\}$ to a base $\{\alpha_1, \ldots, \alpha_n\}$ of $T^*\mathbb{R}^n$. So we can rewrite ω as a linear combination $\sum_{i_1 \leq \ldots \leq i_k} a_{i_1, \ldots, i_k} \alpha_{i_1} \wedge \ldots \wedge \alpha_{i_k}$. If (ii) holds then in each term of the linear combination of ω there must be at least one form α_i with $i \in \{1, \ldots, p\}$. This show that $\omega \in [\alpha_1, \ldots, \alpha_p]$.

(i) $\Rightarrow$ (iii) is obvious.

(iii) $\Rightarrow$ (i) is easily proved by choosing the local coordinate system in such a way that $\xi_i = \frac{\partial}{\partial x^i}$. The conditions $\alpha_i(\xi_j) = 0$ or $\omega(\xi_j) = 0$ then mean that all terms in α_i or ω contain dx^i, and the result is obvious.

We finish this section introducing the definition of *differential ideal*, which is an ideal closed with respect to the differentiation.

DEFINITION 3.6. *Let* I *an ideal. We say that is a* differential ideal *if the following property holds:*

$$\omega \in I \Longrightarrow d\omega \in I.$$

It is easy to see that if $I = [\alpha_1, \ldots, \alpha_p]$ is a differential ideal if and only if there exists forms β_{ij} such that

$$d\alpha_i = \sum_j \beta_{ij} \wedge \alpha_j.$$

An important class of differential ideals is given as follows. Let M be a submanifold of $\mathbb{R}^n$, let $i_M : M \to \mathbb{R}^n$ be the canonical embedding and i_M^* its pullback. Define:

$$I_M := \{\alpha : i_M^* \alpha = 0\}.$$

It turns out by naturality that I_M is a differential ideal. Indeed if $\alpha \in I_M$ and β is any differential form, we have

$$i_M^* (\beta \wedge \alpha) = (i_M^* \beta) \wedge (i_M^* \alpha) = 0,$$
$$i_M^* (d\alpha) = d(i_M^* \alpha) = 0.$$

CHAPTER 4

The Frobenius theorem

4.1. – Statement and proof

Let ω be a differential form on $\mathbb{R}^n$ of class C^∞. We can now try to solve our first problem, which is to find functions λ and V on $\mathbb{R}^n$ such that $\omega = \lambda dV$. Suppose for the moment that the problem is solved, and let us investigate the consequences. The first one is that the manifolds $\{V = \text{constant}\}$ are interesting. Indeed if we pose

$$M(h) := \{x : V(x) = h\},$$

then we have $i^*_{M(h)}dV = 0$, hence $i^*_{M(h)}\omega = 0$ so $i^*_{M(h)}d\omega = 0$. In other words if $E := T_p M(h)$ then $i_E\omega = 0$ and $i_E d\omega = 0$. $M(h)$ is $(n-1)$-dimensional so that by (iii) of Proposition 3.1 there exists some 1-form α such that $d\omega = \alpha \wedge \omega$.

Another way to see this fact is to take the differential of $\omega = \lambda dV$. In this case we obtain

$$d\omega = \frac{d\lambda}{\lambda} \wedge \omega,$$

so that $\alpha = \frac{d\lambda}{\lambda}$. The condition

$$d\omega = \alpha \wedge \omega$$

is called the *Frobenius condition*. It is satisfied if and only if ω generates a differential ideal. It can also be written:

$$\omega \wedge d\omega = 0.$$

The relevance of this condition is explained by the following

THEOREM 4.1   (Frobenius). *Let ω be a 1-form which satisfies $\omega \neq 0$ and $\omega \wedge d\omega = 0$ near some point P. Then there exists two functions λ and V such that*

$$\omega = \lambda dV$$

in a neighbourhood of P_0.

We will deduce the proof of the Frobenius theorem to a more general fact, which we now state as a

LEMMA 4.1.   *Let $\omega_{r+1}, \ldots, \omega_n$ be $n - r$ linearly independent 1-forms. Assume that $[\omega_{r+1}, \ldots, \omega_n]$ generate a differential ideal or, in other words, that $\omega_{r+1}, \ldots, \omega_n$ satisfies the generalized Frobenius condition:*

(4) $$\forall i \quad \exists \alpha_{ij}, \quad s.t. \quad d\omega_i = \sum_j \alpha_{ij} \wedge \omega_j.$$

Then it is possible to find a system of local coordinates $(y_1, \ldots, y_n)$ in such a way that the dy_i and the ω_i, for $i > r$, generate the same differential ideal:

$$[\omega_{r+1}, \ldots, \omega_n] = [dy_{r+1}, \ldots, dy_n].$$

PROOF. Let us consider first the case $r = 1$. This means that we have $n - 1$ forms $\omega_2, \ldots, \omega_n$ on $\mathbb{R}^n$. Incidentally we note that in this case the generalized Frobenius condition (4) is always satisfied. Indeed, completing $\omega_2, \ldots, \omega_n$ to a base by adding an ω_1 we can decompose $d\omega_i$ according to that base:

$$d\omega_i = \sum_{i,j=1}^{n} a_{ij} \omega_i \wedge \omega_j,$$

for some a_{ij} with $a_{ii} = 0$. Clearly, for $i > 1$, every term on the right contains some ω_j with $j > 1$.

To construct the local coordinate system we ask for, we fix a point P_0 and we associate with every P in a neighbourhood of P_0 the kernel of the ω_i:

$$E(P) := \{\xi \in T_P\mathbb{R}^n : \omega_i(\xi) = 0, \ i = 2, \ldots, n\},$$

It follows frome the assumptions that $E(P)$ is a one dimensional subspace, spanned by a tangent vector $\xi(P)$. So we can define a C^∞ vector field $\xi(P)$ in a neighbourhood of P_0, to which is associated a C^∞ flow.

It is well known that there is a local coordinate system $(y_1, \ldots, y_n)$ such that $\xi = \frac{\partial}{\partial y_1}$. This is done by solving the Cauchy problem

$$\begin{cases} \dfrac{d}{dt}\gamma(t) = \xi(\gamma(t)), \\[2mm] \gamma(0) = P_0. \end{cases}$$

and by posing

$$y_1(\gamma(t)) = t, \quad y_i(\gamma(t)) = 0, \quad i = 2, \ldots, n.$$

By construction, the families $\omega_i(P)$ and $dy_i(P)$, for $i > 1$, have the same kernel $\xi(P)$. This means that they generate the same $(n-1)$-dimensional subspace, and hence the same differential ideal (remember that the generalized Frobenius condition is automatically satisfied). Hence the result for $r = 1$.

Now we give the argument for $r = 2$. In this case $I = [\omega_3, \ldots, \omega_n]$. Let us fix a local coordinate system $x_1, \ldots, x_n$ in such a way that dx_1 and dx_2 are linearly independent of $\omega_3, \ldots, \omega_n$ and let J be the subspace spanned by $(dx_2, \omega_3, \ldots, \omega_n)$. Again, since J is $(n-1)$-dimensional, it is also a differential ideal, and we write $J = [dx_2, \omega_3, \ldots, \omega_n]$.

By the first part (induction hypothesis) there exists a coordinate system $y_1, \ldots, y_n$ such that $J = [dy_2, \ldots, dy_n]$. Since the dy_k generate J, there will be some $\beta_2, \ldots, \beta_n$ such that

$$dx_2 = \sum_{k=2}^{n} \beta_k dy_k.$$

At least one of the β_k is not zero. We can suppose that it is β_2. From this it follows that $dx_2, dy_3, \ldots, dy_n$ are linearly independent and that

$$[dy_2, \ldots, dy_n] = [dx_2, dy_3, \ldots, dy_n],$$

Hence we can find, for each $k \geq 3$, some $p_k(y_1, \ldots, y_n)$ and some $\omega'_k \in I$ such that

(5) $$dy_k = p_k dx_2 + \omega'_k,$$

We claim that $I = [\omega'_3, \ldots, \omega'_n]$. Clearly $[\omega'_3, \ldots, \omega'_n] \subset I$. If we show that $\omega'_3, \ldots, \omega'_n$ are linearly independent the equality follows. Suppose that there exists $q_3, \ldots, q_n$ such that

$$\sum_{k=3}^{n} q_k \omega'_k = 0.$$

Then

$$\sum_{k=3}^{n} q_k dy_k - \left(\sum_{k=3}^{n} q_k p_k \right) dx_2 = 0.$$

But the forms $dx_2, dy_3, \ldots, dy_n$ are linearly independent (by the preceding discussion) so that $q_k = 0$ for all $k \geq 3$, as desired.

Now differentiating (5) and recalling that I is a differential ideal we find $dp_k \wedge dx_2 = -d\omega_k' \in I$, or explicitly

$$\sum_{i} \frac{\partial p_k}{\partial y_i} dy_i \wedge dx_2 \in I.$$

Hence $\frac{\partial p_k}{\partial y_1} dy_1 \wedge dx_2 \in I$ and, by our choice of dx_2 and dy_1, this is possible only if

$$\frac{\partial p_k}{\partial y_1} = 0, \quad k \geq 3.$$

Going back to (5) we can say that we have found a local coordinate system such that the forms $\{\omega_k'\}_{3 \leq k \leq n}$ depend only on the coordinates $y_2, \ldots, y_n$. So we are reduced to the first case, and this finishes the proof. □

4.2. – Application to the one consumer model

As an application of the Frobenius theorem, we shall prove the so-called *Slutsky conditions*, i.e. Theorem 2.1. Recall from the one-consumer model that we want to determine when a given function $p \mapsto x(p)$ from $\mathbb{R}^n$ into itself, satisfying $x(p) \cdot p = 1 \ \forall p$ is a *demand function*. This can formally written as follows. Introduce the 1-form ω given by:

$$\omega(p) := \sum_{i} x_i(p) dp_i$$

and find functions $\lambda(p)$ and $V(p)$ satisfying

(6)
$$\begin{cases} x(p) \cdot p = 1, \\ \omega := \sum_{i} x_i(p) dp_i = \lambda dV. \end{cases}$$

In addition, we should require that λ is positive and V quasi-convex, but we are not going into that problem for the moment. The Frobenius condition is then necessary and sufficient; let us find out what it looks like.

To this purpose we introduce the vector field

$$P := \sum_i p_i \frac{\partial}{\partial p_i}.$$

Then the first condition in (6) becomes $\omega(P) = 1$, and the Frobenius condition gives

$$d\omega(P, \xi) = \alpha(P)\omega(\xi) - \alpha(\xi)\omega(P) = \alpha(P)\omega(\xi) - \alpha(\xi),$$

for any vector field ξ. From this we deduce a formula for α:

$$\alpha(\xi) = -d\omega(P, \xi) + \alpha(P)\omega(\xi).$$

In the product $\alpha \wedge \omega$ the last term will disappear. For this reason we can choose

$$\alpha_0(\xi) = -d\omega(P, \xi).$$

Now we have to check that

(7)
$$d\omega(\xi, \eta) = \alpha_0 \wedge \omega(\xi, \eta),$$

for any vector fields ξ, η. It is already the case when $\xi = P$ and η is arbitrary, by our definition of α_0. For arbitrary ξ we have:

$$\alpha_0 \wedge \omega(\xi, \eta) = -d\omega(P, \xi)\omega(\eta) + d\omega(P, \eta)\omega(\xi).$$

Note that

$$d\omega = \sum_{i,j} \frac{\partial x_i}{\partial p_j} dp_j \wedge dp_i.$$

Then

$$d\omega(P, \xi) = \sum_{i,j} \frac{\partial x_i}{\partial p_j} dp_j \wedge dp_i(P, \xi) = \sum_{i,j} \frac{\partial x_i}{\partial p_j}(p_j\xi^i - \xi^j p_i).$$

Rearranging indexes we can write:

$$\alpha_0 \wedge \omega(\xi, \eta) = \sum \left(\frac{\partial x_i}{\partial p_k} p_k x_j + \frac{\partial x_k}{\partial p_j} p_k x_i - \frac{\partial x_j}{\partial p_k} p_k x_i - \frac{\partial x_k}{\partial p_i} p_k x_j \right) \xi^j \eta^i.$$

By the first condition in (6) we have that

$$\frac{\partial x_k}{\partial p_j} p_k = \frac{\partial}{\partial p_j}(x_k p_k) - x_k \frac{\partial p_k}{\partial p_j} = -x_k \delta_j^k = -x_j,$$

so that finally we arrive to

$$\alpha_0 \wedge \omega(\xi, \eta) = \sum_k \left(\frac{\partial x_i}{\partial p_k} p_k x_j - \frac{\partial x_j}{\partial p_k} p_k x_i \right) \xi^j \eta^i.$$

On the other hand

$$d\omega(\xi, \eta) = \frac{\partial x_i}{\partial p_j} dp_j \wedge dp_i(\xi, \eta)$$

$$= \frac{\partial x_i}{\partial p_j} \left(dp_j(\xi) dp_i(\eta) - dp_j(\eta) dp_i(\xi) \right)$$

$$= \frac{\partial x_i}{\partial p_j} \left(\xi^j \eta^i - \eta^j \xi^i \right)$$

$$= \left(\frac{\partial x_i}{\partial p_j} - \frac{\partial x_j}{\partial p_i} \right) \xi^j \eta^i.$$

Equality (7) is true if and only if

$$\frac{\partial x_i}{\partial p_j} - \sum_k \frac{\partial x_i}{\partial p_k} p_k x_j = \frac{\partial x_j}{\partial p_i} - \sum_k \frac{\partial x_j}{\partial p_k} p_k x_i,$$

or, equivalently, if and only if the matrix $(\sigma_{ij})_{i,j}$ is symmetric, where

$$\sigma_{ij} = \left(\frac{\partial x_i}{\partial p_j} - \sum_k \frac{\partial x_i}{\partial p_k} p_k x_j \right).$$

This is just the Slutsky conditions, as stated in Section 2.

CHAPTER 5

The household model

5.1. – Statement of the model and Pareto optimality

Up to now we have considered the simplest model, namely a situation where there is only one consumer. We have deduced the Slutsky conditions for this case which are necessary and sufficient to rebuild the utility functions. In this section we discuss what happens if we try to study more complex systems.

Let us suppose that there are $n + N$ goods and two consumers (one can easily imagine that the situation can be more complicated, nevertheless what we consider here contains all the obstructions which appear in more general situations). Our way to denote the number of goods is due to the fact that we denote by n the number of *private goods* and by N the number of *public goods* within the household. Food and clothes are private goods, because they are consumed singly; lodging and TV are public goods, because they can be enjoyed simultaneously by both partners.

Each consumer is characterized by his own individual utility function:

$$U_i : \mathbb{R}^n \times \mathbb{R}^n \times \mathbb{R}^N \to \mathbb{R}$$

for $i = 1, 2$. We will denote by $x_1 \in \mathbb{R}^n$ the private consumption of the first consumer, by $x_2 \in \mathbb{R}^n$ the private consumption of the second one and by $X \in \mathbb{R}^n$ the common consumption of public goods. Note that the utility function of a consumer depends not only on his own consumption, but also on the private consumption of the other consumer. If one partner smokes, it may inconvenience the other; on the other hand, if one partner plays an instrument, the other one may enjoy the sound. These are called externalities.

Of course various restrictive assumptions can be made: *no external-ities* (in this case $U_i = U_i(x_i, X)$), *no public goods* ($U_i = U_i(x_1, x_2)$), *no private goods* ($U_i = U_i(X)$). This problem is called the *household problem*, because it is typical of two people living in the same house, and pooling their resources, which they then have to share between their private and public consumptions.

This problem differs from the preceding one in the fact that there are now two criterions to be optimized simultaneously. Multicriterion optimization has been well developed for the purpose of economics, and the appropriate definition is that of Pareto optimality, to which we now proceed.

Denote by p the price of private goods, and by P the price of public goods. For the sake of simplicity, set the wealth of the household equal to 1.

DEFINITION 5.1. We say that an allocation (x_1, x_2, X) is *feasible* iff:

$$p \cdot (x_1 + x_2) + P \cdot X \le 1$$

We say that it is *Pareto optimal* if it is feasible and if there do not exist other feasible allocations which can strictly increase at least one individual's utility U_i without decreasing the utility of the other. In other words,

$$p \cdot (y_1 + y_2) + P \cdot Y \le 1 \qquad \text{and} \qquad (y_1, y_2, Y) \ne (x_1, x_2, X)$$

implies

$$U_1(y_1, y_2, Y) < U_1(x_1, x_2, X) \qquad \text{or} \qquad U_2(y_1, y_2, Y) < U_2(x_1, x_2, X).$$

In other words, if (x_1, x_2, X) is *not* Pareto optimal, then the total wealth of the household can be spent in such a way that both partners are better off. Take the elementary case where $n = 1$ and $U_1(x_1) = x_1$, $U_2(x_2) = x_2$ (no externalities nor public goods). In that case, the problem is simply to split 1 dollar. Then (x_1, x_2) is Pareto optimal iff $x_1 + x_2 = 1$, that is, there is no money left. This gives us the intuitive meaning of Pareto optimality: no resource is wasted.

It can be proved, using the standard separation theorem for convex sets, that (x_1, x_2, X) is Pareto optimal iff there exists some constant $\mu \geq 0$ such that (x_1, x_2, X) maximizes $U_1 + \mu U_2$ over all feasible allocations. It follows that there are infinitely many Pareto optima, one for every choice of μ. In the household problem, the parameter $\mu \in \mathbb{R}_+$ indicates the relative power of the two consumers in the allocation process. It will be taken to be a smooth function of the price system (p, P).

The household problem can now be stated as an optimization problem:

$$(\mathcal{P}_{p,P,\mu,U_1,U_2}) \begin{cases} \max \{U_1(x_1, x_2, X) + \mu(p, P)U_2(x_1, x_2, X)\}), \\[2mm] x_1, x_2 \in \mathbb{R}^n_+, \quad X \in \mathbb{R}^N_+, \\[2mm] p \cdot (x_1 + x_2) + P \cdot X \leq 1. \end{cases}$$

As in the case of the single consumer we introduce the *first order necessary condition* (in the sequel (FONC)) for optimality.

DEFINITION 5.2. Let Θ an open subset of $\mathbb{R}^n \times \mathbb{R}^N$, $(\hat{x}_1, \hat{x}_2, \hat{X})$: $\Theta \to \mathbb{R}^n_+ \times \mathbb{R}^n_+ \times \mathbb{R}^N_+$ a smooth mapping. We shall say that

$$(\hat{x}_1, \hat{x}_2, \hat{X}) \in \text{FONC}\,(U_1, U_2, \mu, \Theta),$$

if there exists a smooth function $\lambda : \Theta \to \mathbb{R}_+$ such that

(i) $$p \cdot (\hat{x}_1(p, P) + \hat{x}_2(p, P)) + P \cdot \hat{X}(p, P) = 1,$$

(ii) $$\frac{\partial}{\partial x^i_k} [U_1 + \mu(p, P)U_2]\Big|_{(\hat{x}_1, \hat{x}_2, \hat{X})} = \lambda(p, P)p_i, \quad i = 1, \ldots, n, \; k = 1, 2$$

(iii) $$\frac{\partial}{\partial X^j} [U_1 + \mu(p, P)U_2]\Big|_{(\hat{x}_1, \hat{x}_2, \hat{X})} = \lambda(p, P)P_j, \quad j = 1, \ldots, N,$$

(iv) $$\hat{x}^i_k(p, P) > 0, \; \hat{X}^j(p, P) > 0, \; \lambda(p, P) \neq 0.$$

It is well known that if $(\bar{x}_1, \bar{x}_2, \bar{X})$ solves the problem $(\mathcal{P}_{p,P,\mu,U_1,U_2})$ for the particular value $(\bar{p}, \bar{P}) \in \mathbb{R}^n_+ \times \mathbb{R}^N_+$, then there will usually exists some neighbourhood Θ of $(\bar{p}, \bar{P})$ and a function $(\hat{x}_1, \hat{x}_2, \hat{X}) \in \text{FONC}(U_1, U_2, \mu, \Theta)$ with $\hat{x}_i(\bar{p}, \bar{P}) = \bar{x}_i$ and $\hat{X}(\bar{p}, \bar{P}) = \bar{X}$. "Usually" means here that some non-degeneracy conditions, involving first and second derivatives of U_1 and U_2 at the point $(\bar{x}_1, \bar{x}_2, \bar{X})$ have to hold: we shall not go into the details. It is also the case that FONC are not sufficient, unless U_1 and U_2 are concave (and μ positive).

5.2. – The Browning-Chiappori necessary conditions

As in the single consumer case, we want to characterize demand functions which arise from the problem $(\mathcal{P}_{p,P,\mu,U_1,U_2})$. But this time we cannot observe the individual consumptions of private goods $\hat{x}_1$ and $\hat{x}_2$. Only the aggregate demand function of the household

$$\hat{x}(p, P) := \hat{x}_1(p, P) + \hat{x}_2(p, P),$$

can be observed together with $\hat{X}(p, P)$, the demand for public goods. Browning and Chiappori [3] have found a necessary condition for $(\hat{x}, \hat{X})$ to be a household demand function. They have proved the following

PROPOSITION 5.1. *If there exists* U_1, U_2, μ, Θ *and* $\hat{x}_1, \hat{x}_2$ *in such a way that* $(\hat{x}_1, \hat{x}_2, \hat{X}) \in$ FONC (U_1, U_2, μ, Θ) *and* $\hat{x} = \hat{x}_1 + \hat{x}_2$, *then the Slutsky matrix associated to* $(\hat{x}, \hat{X})$ *is a sum of a symmetric matrix and a matrix with rank one.*

To write down explicitly the conclusion of the Proposition 5.1 let us denote by $\pi := (p, P) \in \mathbb{R}^{n+N}$ and $\xi := (\hat{x}, \hat{X})$. Recall that the Slutsky matrix is given by

$$\sigma^{ij} = \frac{\partial \xi^i}{\partial \pi_j} - \sum_k p_k \frac{\partial \xi^i}{\partial p_k} \xi^j.$$

So the Browning-Chiappori result states that

$$\sigma^{ij} = s^{ij} + a^i b^j, \quad \text{with } s^{ij} = s^{ji}.$$

Browning and Chiappori also raise the question whether this condition is sufficient. We will give a positive answer by using exterior differential calculus (and simultaneously give a direct proof that the above condition is necessary).

In order to approach the problem we have presented in the previous section we follow the same procedure as in the single consumer case, namely we translate the problem into the language of differential forms. The first step is given by the following

LEMMA 5.1. *Let* $\xi : \Theta \to \mathbb{R}^n_+ \times \mathbb{R}^N_+$ *be given. If there are smooth functions* $(\hat{x}_1, \hat{x}_2)$ *and* (U_1, U_2, μ) *such that*

(8) $\xi = (\hat{x}_1 + \hat{x}_2, \hat{X}), \quad (\hat{x}_1, \hat{x}_2, \hat{X}) \in FONC(U_1, U_2, \mu, \Theta),$

then there exist smooth functions φ, ψ, f and g on Θ such that

(9)
$$\sum_{i=1}^{n+N} \xi^i(\pi) d\pi_i = \varphi df + \psi dg.$$

Conversely, if there exists an open subset $\mathcal{U} \subset \mathbb{R}_+^l$ such that $\xi : \Theta \to \mathcal{U}$ is smoothly invertible, if (9) holds for some smooth functions (φ, ψ, f, g) and if $\pi \cdot \xi(\pi) = 1$ on Θ, then, setting $n = 0$, $N = l$ and $\hat{X} = \xi$ (no private goods), smooth functions (U_1, U_2) can be found in such a way that (8) holds.

PROOF. Suppose that (8) holds. Let us introduce the indirect utility functions:

$$\hat{U}_i(p, P) := U_i(\hat{x}_1(p, P), \hat{x}_2(p, P), \hat{X}(p, P)), \quad i = 1, 2,$$

and

$$\hat{U}(p, P) = \hat{U}_1(p, P) + \mu(p, P)\hat{U}_2(p, P)$$

$$= \max_{x_1, x_2, X} \left\{ \begin{array}{l} U_1(x_1, x_2, X) + \mu(p, P)U_2(x_1, x_2, X) \\ + \lambda(p, P)\left[1 - p \cdot (x_1 + x_2) - P \cdot X\right] \end{array} \right\}.$$

By the envelope theorem

$$\frac{\partial \hat{U}}{\partial \pi_i} = \frac{\partial \mu}{\partial \pi_i} \hat{U}_2 - \lambda \xi^i,$$

which we rewrite as

$$\xi^i = \frac{\hat{U}_2}{\lambda} \frac{\partial \mu}{\partial \pi_i} - \frac{1}{\lambda} \frac{\partial \hat{U}}{\partial \pi_i}.$$

Hence we deduce

$$\sum_{i=1}^{n+N} \xi^i d\pi_i = \frac{\hat{U}_2}{\lambda} d\mu - \frac{1}{\lambda} d\hat{U},$$

as announced. Note that in particular we have shown that $\psi = -\lambda^{-1}$, $g = \hat{U}$, $-\varphi = \hat{U}_2 \psi$ and $f = \mu$ as a solution (not unique).

For the converse let us suppose that (9) holds for some functions (φ, ψ, f, g) and $\pi \cdot \xi(\pi) = 1$ on Θ. This last condition implies that

ξ cannot vanish, hence we can assume that ψ does not vanish on Θ (eventually restricting Θ). Now we define

$$\lambda := -\frac{1}{\psi}, \quad \hat{U} := g, \quad \hat{U}_2 := -\frac{\varphi}{\psi}, \qquad \mu := f, \quad \hat{U}_1 := \hat{U} - \mu\hat{U}_2,$$

and

$$U_1(X) := \hat{U}_1(\xi^{-1}(X)), \quad U_2(X) := \hat{U}_2(\xi^{-1}(X)).$$

Finally we can pass to verify conditions (8). The first one, in our setting, is trivial. Let us check only the second one, which means to check conditions (i)-(iv) stated in the Definition 5.2. In our setting (i) and (iv) are immediate. It remains only (ii), (iii). Note that by $\pi \cdot \xi(\pi) = 1$ and our definitions,

$$\hat{U}(\pi) = U_1(\xi(\pi)) + \mu(\pi)U_2(\xi(\pi)) + \lambda(\pi)(1 - \pi \cdot \xi(\pi))$$

$$= \hat{U}_1(\pi) + \mu(\pi)\hat{U}_2(\pi) + \lambda(\pi)(1 - \pi \cdot \xi(\pi)).$$

Differentiating we get

$$(10) \quad d\hat{U} = d\hat{U}_1 + \mu d\hat{U}_2 + \hat{U}_2 d\mu - \lambda\xi^i d\pi_i - \lambda\pi_i \frac{\partial\xi^i}{\partial\pi_j} d\pi_j + (1 - \pi\cdot\xi(\pi))d\lambda.$$

But the last term in the previous equality is equal to zero due to the condition $\pi \cdot \xi(\pi) = 1$. On the other hand writing down (9) in our setting we deduce

$$(11) \qquad d\hat{U} = dg = \frac{1}{\psi}\left(\xi^i d\pi_i - \varphi df\right) = -\lambda\xi^i d\pi_i + \hat{U}_2 d\mu.$$

Comparing (10) with (11) we deduce the following equation

$$(12) \qquad\qquad d\hat{U}_1 + \mu d\hat{U}_2 - \lambda\frac{\partial\xi^i}{\partial\pi_j}\pi_i d\pi_j = 0.$$

Inverting the (12) it is easy to find conditions (ii) and (iii) of the definition 5.2. $\square$

REMARK 5.1. Lemma 5.1 states that if the Browning-Chiappori condition is satisfied, then the problem has a solution with public goods only. If we know in advance that there are really private goods, then the Browning-Chiappori condition is no longer sufficient, and addi-tional conditions must be satisfied. They will be described in the next section.   □

We now have to solve the following problem: given a 1-form ω find functions (φ, ψ, f, g) in such a way that $\omega = \varphi df + \psi dg$.

This is a generalization of the problem that we have posed previously and that we have solved via the Frobenius theorem. Let us first look for necessary condition on the form ω. Differentiating the equation satisfied by ω we obtain $d\omega = d\varphi \wedge df + d\psi \wedge dg$. Hence

$$d\omega \wedge d\omega = 2d\varphi \wedge df \wedge d\psi \wedge dg.$$

Multiplying once again by ω we find the following necessary condition:

(13) $\omega \wedge d\omega \wedge d\omega = 0.$

It is possible to give other conditions, as in the following

PROPOSITION 5.2. *Let ω be a differential form. Each of the following conditions is equivalent to* (13):

(a) *there exists a 3-form α such that $d\omega \wedge d\omega = \alpha \wedge \omega$;*
(b) *there exists a 2-form β and a 1-form γ such that $d\omega = \beta + \gamma \wedge \omega$ and $\beta \wedge \beta = 0$.*

PROOF. Let us construct a base of 1-forms, let say $\omega_1, \ldots, \omega_n$ with $\omega_1 = \omega$. Then

$$d\omega = \sum_{ij} a_{ij}\omega_i \wedge \omega_j = \omega_1 \wedge \gamma + \beta,$$

where β contains no term of the form $\omega_1 \wedge \omega_j$. From this the equivalence between (a) and (b) follows immediately. Indeed if (a) holds we have

$$\alpha \wedge \omega = d\omega \wedge d\omega = \beta \wedge \beta + 2\omega_1 \wedge \gamma \wedge \beta,$$

and by construction β does not contain $\omega = \omega_1$ so $\beta \wedge \beta$ has to vanish. On the other hand if (b) holds (a) follows immediately.

It is also clear that (a) implies (13). Conversely, if (13) is satisfied, we may again use the decompostion $d\omega = \omega_1 \wedge \gamma + \beta$; writing this in (13) with $\omega_1 = \omega$, we get $\omega \wedge \beta \wedge \beta = 0$, and since β does not contain ω, this leads to $\beta \wedge \beta = 0$.   □

CHAPTER 6

Pfaff's theorem

In this section we will finally prove that the Browning-Chiappori condition is necessary and sufficient (in the case of public goods). This is done by applying the Pfaff theorem, which solves (among others) the following problem: given a 1-form ω, when can we find functions (φ, ψ, f, g) satisfying $\omega = \varphi df + \psi dg$?

THEOREM 6.1 (Pfaff). *Let ω be a differential form. Suppose that there exists an open set $\mathcal{U} \subset \mathbb{R}^n$ such that*

$$(14) \qquad\qquad \omega \wedge d\omega \neq 0 \ in \, \mathcal{U},$$

$$(15) \qquad\qquad \omega \wedge d\omega \wedge d\omega = 0 \ in \, \mathcal{U}.$$

Then there exists an open subset $\mathcal{V} \subset \mathcal{U}$, functions $\varphi, \psi, f, g \in C^\infty(\mathcal{V}; \mathbb{R})$ such that

$$\omega = \varphi df + \psi dg.$$

REMARK 6.1. Note that the Pfaff and the Frobenius theorems have conditions on the form ω which exclude the case of a form ω that satisfies both of them. The Pfaff theorem tells us when there exists a two-terms decompostion $\omega = \varphi df + \psi dg$, while the Frobenius theorem gives us a single one $\omega = \varphi df$. This is interesting in applications, because we could have to test a demand function of unknown type: we might not know if we are dealing with a single consumer or a household.

If the Frobenius condition does not hold but the Pfaff condition does, then we have in hand the demand function of a household. □

PROOF. FIRST STEP: Suppose for a moment that $\omega = \varphi df + \psi dg$. Then $d\omega = d\varphi \wedge df + d\psi \wedge dg$. So, the first step will be to search a decomposition for $d\omega$ of the type

$$d\omega = \omega \wedge \omega' + \gamma \wedge \gamma'.$$

By Proposition 5.2 we know that $d\omega$ has the form

$$d\omega = \omega \wedge \omega' + \sigma,$$

and we have the information that $\sigma \wedge \sigma = 0$. We need from this a factorization for σ. This purpose comes from the following

LEMMA 6.1. *Let σ a 2-form, $\sigma \neq 0$ such that*

$$\sigma \wedge \sigma = 0.$$

Then there exist γ, γ' 1-forms such that $\sigma = \gamma \wedge \gamma'$.

PROOF OF LEMMA 6.1. Let us introduce the one-form

$$u(\xi)(\eta) = \sigma(\xi, \eta).$$

Because $\sigma \neq 0$ there exists a ξ such that $u(\xi) \neq 0$. Now it turns out that

$$u(\xi) \wedge \sigma = 0.$$

Indeed, we have:

$$(u(\xi) \wedge \sigma)(\eta_1, \eta_2, \eta_3)$$
$$= 2[\sigma(\xi, \eta_1)\sigma(\eta_2, \eta_3) + \sigma(\xi, \eta_2)\sigma(\eta_3, \eta_1) + \sigma(\xi, \eta_3)\sigma(\eta_1, \eta_2)]$$
$$= (\sigma \wedge \sigma)(\xi, \eta_1, \eta_2, \eta_3) = 0.$$

Our claim now is that $\sigma = u(\xi) \wedge \gamma'$. Let $\alpha_1 := u(\xi)$ and complete this form to a base $\alpha_1, \alpha_2, \ldots, \alpha_n$. Then

$$\sigma = \sigma^{ij}\alpha_i \wedge \alpha_j.$$

From $\alpha_1 \wedge \sigma = 0$, it follows that

$$\sum_{i,j \neq 1} \sigma^{ij}\alpha_i \wedge \alpha_j \wedge \alpha_1 = 0,$$

from which it follows that $\sigma^{ij} = 0$ for all $i, j > 1$. Hence

$$\sigma = \sum_{i=1 \vee j=1} \sigma^{ij} \alpha_i \wedge \alpha_j = \alpha_1 \wedge \gamma',$$

for some appropriate γ'.   □

SECOND STEP: By the first step $d\omega = \omega \wedge \omega' + \gamma \wedge \gamma'$. Note that if γ or γ' are null forms, we have just the Frobenius condition. Nevertheless, as we have seen in the Lemma 6.1, γ is not 0 if σ is not, and in our setting this is implied by (14). The purpose of this second step is to show that $[\omega, \gamma, \gamma'] = [dy_1, dy_2, dy_3]$, and to apply the Frobenius theorem. We first have to show that the linear span of $(\omega, \gamma, \gamma')$ generates a differential ideal. We claim that:

$$(16) \qquad \mathrm{Span}(\omega, \gamma, \gamma') = \{\alpha : \alpha \wedge \omega \wedge d\omega = 0\} =: J.$$

Let us prove that the right hand side is a differential ideal. Let $\alpha \in J$, then

$$0 = d\,[\alpha \wedge \omega \wedge d\omega]$$

$$= d\alpha \wedge \omega \wedge d\omega + \alpha \wedge d\,[\omega \wedge d\omega]$$

$$= d\alpha \wedge \omega \wedge d\omega + \alpha \wedge d\omega \wedge d\omega.$$

On the other hand, we have:

$$d\omega \wedge d\omega = 2\omega \wedge \omega' \wedge \gamma \wedge \gamma' = -\omega' \wedge \omega \wedge d\omega.$$

Hence

$$\alpha \wedge d\omega \wedge d\omega = \omega' \wedge \alpha \wedge \omega \wedge d\omega = 0,$$

from which follows that $d\alpha \wedge \omega \wedge d\omega = 0$, namely J is a differential ideal.

Now we prove (16). It is easy to verify that $\mathrm{Span}(\omega, \gamma, \gamma') \subset J$. For the converse let $\alpha \in J$ be a 1-form (the general case follows in the same way). Because ω, γ, γ' are linearly independent (by construction) we can complete them into a base $\alpha_1, \ldots, \alpha_n$ in such a way that $\alpha_1 = \omega$, $\alpha_2 = \gamma$ and $\alpha_3 = \gamma'$. Let us write α as linear combination of these forms, then take the external product with $\omega \wedge d\omega$. We obtain

$$0 = \alpha \wedge \omega \wedge d\omega = \left(\sigma^i \alpha_i\right) \wedge \omega \wedge \gamma \wedge \gamma'$$

$$= \sum_{i=4}^{n} \sigma^i \alpha_i \wedge \alpha_1 \wedge \alpha_2 \wedge \alpha_3.$$

By this follows that $\sigma^i = 0$ for $i = 4, \ldots, n$, so that

$$\alpha = \sigma^1\omega + \sigma^2\gamma + \sigma^3\gamma'.$$

We have proved that $\text{Span}(\omega, \gamma, \gamma')$ is a differential ideal. By the Frobenius theorem it follows that, for an appropriate coordinate system $(y_1, \ldots, y_n)$, we have:

$$[\omega, \gamma, \gamma'] = [dy_1, dy_2, dy_3].$$

THIRD STEP: By the preceding step, there is a decomposition of ω into a sum of three terms $\omega = \lambda_1 dy_1 + \lambda_2 dy_2 + \lambda_3 dy_3$. We want to reduce this decomposition to two terms. Let us suppose that $\lambda_1 \neq 0$, and consider the ideal generated by ω and dy_1. As before, we show that:

$$[\omega, dy_1] = \{\alpha : \alpha \wedge \omega \wedge dy_1 = 0\} =: J',$$

and J' is a differential ideal. Using the Frobenius theorem again, we have $J = [dz_1, dz_2]$, hence the conclusion.   □

REMARK 6.2. Note that if $n \leq 4$, $\omega \wedge d\omega \wedge d\omega$ as a 5-form is identically zero. So in this case the only condition on ω is (14).   □

More generally the following theorem holds

THEOREM 6.2. *Let ω a differential form such that there exists an open set $\mathcal{U} \subset \mathbb{R}^n$ such that*

$$\omega \wedge \bigwedge_{i=1}^{k-1} d\omega \neq 0 \ on\,\mathcal{U},$$

and

$$\omega \wedge \bigwedge_{i=1}^{k} d\omega = 0 \ on\,\mathcal{U}.$$

Then for any point $\bar{x}$ in $\mathcal{U}$, there exists functions λ_i, V_i, $i = 1, \ldots, k$, defined near $\bar{x}$, and such that

$$\omega = \sum_{i=1}^{k} \lambda_i dV_i.$$

The proof of the Theorem 6.2 is very similar to the one of the Theorem 6.1. We have immediately the

COROLLARY 6.1. *Every regular 1-form on $\mathbb{R}^n$ is a linear combination of at most $\frac{n}{2}$ gradients if n is even, $\frac{n+1}{2}$ if n is odd.*

It is now immediate to see that the Browning-Chiappori conditions are sufficient to reconstruct the utility functions of a given household policy ξ. Indeed, define the linear form ω by $\omega(\pi) = \sum \xi^i(\pi)d\pi_i$. Write down the Browning-Chiappori condition:

$$\sum \sigma^{ij}d\pi_i \wedge d\pi_j = \sum \left(\frac{\partial \xi^i}{\partial \pi_j} - \sum \pi_k \frac{\partial \xi^i}{\partial \pi_k}\xi^j \right) d\pi_i \wedge d\pi_j$$

$$= \sum a^i b^j d\pi_i \wedge d\pi_j .$$

Rewrite it as follows:

(17) $$d\omega = a \wedge b + \gamma \wedge \omega ,$$

with $a = \sum a^i d\pi_i$, $b = \sum b^j d\pi_j$, $\gamma = \sum \pi_k \frac{\partial \xi^i}{\partial \pi_k}d\pi_i$. Clearly, relation (17) implies that $\omega \wedge d\omega \wedge d\omega = 0$, so the Pfaff condition holds, and Lemma 5.1 applies.

REMARK 6.3. There is an additional question which we have not touched, and it is the question of concavity. In microeconomic models, like the Browing-Chiappori one, it is not enough to find fuctions U_1 and U_2 satisfying the first-order necessary conditions: one would also like these functions to be concave, so that the necessary conditions are also sufficient, and the problem is completely solved. Mathematically, this translates into the following question: what additional assumptions are necessary in Pfaff's theorem so that the functions f and g are concave, and the functions ϕ and ψ positive? After these lectures notes were complete, this question was answered in [10].   □

Up to now, we assumed that there are only public goods in the household. In other words, the utility functions are $U_1(X)$ and $U_2(X)$. In this section we deal with the opposite situation, when there are only private goods. In other words, the utility functions are $U_1(x_1)$ and $U_2(x_2)$. The household problem then is as follows:

$$(\mathcal{P}_{p,\mu,U_1,U_2}) \begin{cases} \max \{U_1(x_1) + \mu(p)U_2(x_2)\}, \\ \\ p \cdot (x_1 + x_2) \le 1. \end{cases}$$

Given a function $p \mapsto x(p)$, we want to find functions $x_1, x_2, \mu, U_1,$ U_2 in such a way that $x(p) = x_1(p) + x_2(p)$ and the pair $(x_1(p), x_2(p))$ is the solution of $(\mathcal{P}_{p,\mu,U_1,U_2})$.

Our approach now will be different from the preceding one: we will seek directly $U_1(x)$ and $U_2(x)$, instead of $V_1(p)$ and $V_2(p)$. For this reason, we start from the *inverse demand function* $x \mapsto p(x)$, which is considered given. Note that it can be obtained by simply inverting the usual demand function $p \mapsto x(p)$.

Introducing Lagrange multipliers we get the following optimality conditions

(18)
$$\begin{cases} \dfrac{\partial U_1}{\partial x_1^i}(x_1) = \lambda(x_1 + x_2) p_i(x_1 + x_2), \\[4mm] \dfrac{\partial U_2}{\partial x_2^i}(x_2) = \dfrac{\lambda(x_1 + x_2)}{\mu(x_1 + x_2)} p_i(x_1 + x_2). \end{cases}$$

Note that all the functions here are unknown, with the exception of $p(x)$. Introducing appropriate functions φ_1, φ_2, we can rewrite (18) as follows:

(19)
$$\begin{cases} \varphi_1(x) + \varphi_2(x) = x, \\[3mm] \dfrac{\partial U_1}{\partial x_1^i}(\varphi_1(x)) = \lambda(x) p_i(x), \\[3mm] \dfrac{\partial U_2}{\partial x_2^i}(\varphi_2(x)) = \dfrac{\lambda(x)}{\mu(x)} p_i(x). \end{cases}$$

Let us rewrite (19) in the language of differential form. Setting

$$\alpha_1 := \sum p_i(x) dx_1^i,$$

the first equation in (18) becomes

$$dU_1 = \frac{\partial U_1}{\partial x_1^i}(x_1) dx_1^i = \lambda(x) \alpha_1,$$

while the second one becomes

$$dU_2 = \frac{\partial U_2}{\partial x_2^i}(x_2) dx_2^i = \frac{\lambda(x)}{\mu(x)} \alpha_2,$$

where

$$\alpha_2 := \sum p_i(x) dx_2^i.$$

We now rewrite the full system (19) in a geometrical way. Introduce first the space $\mathbb{R}^{2l+2}$ with coordinates $(x_1^i, x_2^i, \lambda, \mu)$, and in this space introduce the manifold M defined by:

$$(20) \qquad M := \left\{ (x_1^i, x_2^i, \lambda, \mu) : \ x_1 = \varphi_1(x), x_2 = \varphi_2(x) \right\},$$

Denote by i_M^* the pullback operator from $\mathbb{R}^{2l+2}$ to M. System (19) can now be written:

$$(21) \qquad \begin{cases} i_M^* dU_1 = \lambda(x) i_M^* \alpha_1, \\[2mm] i_M^* dU_2 = \dfrac{\lambda(x)}{\mu(x)} i_M^* \alpha_2. \end{cases}$$

So the new problem is to find a submanifold M of $\mathbb{R}^{2l+2}$, and functions λ, μ, U_1, U_2 satisfying the following conditions:

(a) M has dimension l and is the graph of a function of $x_1 + x_2$;
(b) $i_M^* \alpha_1 = \lambda(x)^{-1} i_M^* dU_1$;
(c) $i_M^* \alpha_2 = \mu(x) \lambda(x)^{-1} i_M^* dU_2$.

By the Frobenius theorem conditions (b) and (c) are equivalent to the following ones:

(b') $i_M^* (\alpha_1 \wedge d\alpha_1) = 0$;
(c') $i_M^* (\alpha_2 \wedge d\alpha_2) = 0$.

The Cartan-Kähler theorem, which we shall describe in the next section, enables us to prove the following result (see [6]):

PROPOSITION 6.1. *Assume that there are 1-forms $\beta, \beta', \gamma, \gamma'$ such that:*

$$(22) \qquad \begin{cases} d\alpha_1 \wedge d\alpha_1 = \beta \wedge \alpha_1 \wedge d\alpha_1 + \gamma \wedge \alpha_2 \wedge d\alpha_2, \\[2mm] d\alpha_2 \wedge d\alpha_2 = \beta' \wedge \alpha_1 \wedge d\alpha_1 + \gamma' \wedge \alpha_2 \wedge d\alpha_2. \end{cases}$$

Then we can find U_1, U_2, μ such that $p(x)$ is the inverse demand function corresponding to $\mathcal{P}_{p,\mu,U_1,U_2}$. □

We have now fully solved the household problem (up to concavity). In the case of public goods, the Browning-Chiappori conditions are necessary and sufficient. In the case of private goods, conditions (22) are necessary and sufficient. They imply the Browning-Chiappori ones, but they are stronger. Indeed, take $l = 3$ and set:

$$p_1(x) = \frac{\gamma_1}{x_1}, \quad p_2(x) = \frac{\gamma_2}{x_2}, \quad p_3(x) = \frac{\gamma_3}{x_3}, \quad \text{with } \gamma_1 + \gamma_2 + \gamma_3 = 1,$$

then conditions (22) are not satisfied, but the Browning-Chiappori ones are, as the reader will check.

Note, however, two things. The first one is that the problem is solved only locally, that is, the functions are defined in some unknown neighbourhhod of a prescribed point. The second one is that we have said nothing about the concavity of the functions U_1 and U_2. For the solution to make economic sense, they have to be concave; the result in [10] then gives us additional conditions which have to be met.

We now turn to the general statement of the Cartan-Kähler theorem.

CHAPTER 7

The Cartan-Kähler theorem

We now present the key result upon which our approach relies. This theorem, due to Cartan and Kähler, solves the following, general problem. Given a certain family of differential forms (not necessarily 1-forms, nor even of the same degree), a point $\bar{x}$ and an integer $m \geq 1$, can one find some m-dimensional submanifold M containing $\bar{x}$ and on which all the given forms vanish?

7.1. – An introductory example

As an introduction, let us start from a simple version of our problem, namely the Cauchy-Lipschitz theorem for ordinary differential equations. It states that, given a point $\bar{x} \in \mathbb{R}^n$ and a C^1 function f, defined from some neighborhood $\mathcal{U}$ of $\bar{x}$ into $\mathbb{R}^n$, there exists some $\epsilon > 0$ and a C^1 function $\varphi: \,] - \epsilon, \epsilon \, [\longrightarrow \mathcal{U}$ such that

(23)
$$\begin{cases} \dfrac{d\varphi}{dt} = f(\varphi(t)) & \forall t \in \,] - \epsilon, \epsilon \, [, \\ \\ \varphi(0) = \bar{x}. \end{cases}$$

It follows that $\frac{d\varphi}{dt}(0) = f(\bar{x})$. If $f(\bar{x}) = 0$, the solution is trivial, $\varphi(t) = \bar{x}$ for all t so we assume that $f(\bar{x})$ does not vanish.

This theorem can be rephrased in a geometric way. Consider the graph M of φ:

$$M \; = \; \{\; (t, \varphi(t)) \mid \; -\epsilon < t < \epsilon \;\}.$$

This is a 1-dimensional submanifold of $]-\epsilon, \epsilon[\times \mathcal{U}$. Now, let us introduce the 1-forms ω^i defined by:

$$\omega^i \; = \; f^i(x)dt \; - \; dx^i, \quad 1 \le i \le n-1.$$

Clearly φ solves the differential equation (23) if and only if, for any $x = \varphi(t)$ on M, the ω^i all vanish on $T_x M$, the tangent space to M at x (remember that $T_x M$ is spanned by the vector $v(t) = (1, \frac{d\varphi_1}{dt}, \ldots, \frac{d\varphi_n}{dt})$, and note that $\omega^i[v(t)] = f^i(\varphi(t)) - \frac{d\varphi_i}{dt}$). Equivalently, substituting $x^i = \varphi^i(t)$ into formula (4) yields the pullbacks:

$$\varphi^* \omega^i \; = \; \left[f^i\left(\varphi(t)\right) \; - \; \frac{d\varphi^i}{dt}(t) \right] dt$$

which vanish if and only if φ solves equation (23).

So the Cauchy-Lipschitz theorem tells us how to find a 1-dimensional submanifold of $\mathbb{R} \times \mathbb{R}^n$ on which certain 1-forms vanish.

7.2. – The general problem

The Cauchy-Lipschitz theorem deals with 1-forms of a specific nature. By extension, the fully general problem can formally be stated as follows.

DEFINITION 7.1. Let ω^k, $1 \le k \le K$, be differential forms of degree d_k on an open subset of $\mathbb{R}^n$, and $M \subset \mathbb{R}^n$ a submanifold. We call M an *integral manifold* of the system:

(24) $\omega^1 = 0, \ldots, \omega^K = 0$

if, for any x on M, the ω^i all vanish on $T_x M$, the tangent space to M at x:

(25) $\omega^k \left(\xi^1, \ldots, \xi^{d_k} \right) = 0, \quad 1 \le k \le K,$

whenever $x \in M$ and $\xi^i \in T_x M$ for $1 \leq i \leq d_k$.

An equivalent statement is that the pullbacks $i_M^x(\omega_i)$ of the ω^i to M must all vanish.

Given a point $\bar{x} \in \mathbb{R}^n$, the Cartan-Kähler theorem will give necessary and sufficient conditions for the existence of an integral manifold containing $\bar{x}$.

Necessary conditions are easy to find. Assume an integral manifold $M \ni \bar{x}$ exists, and let m be its dimension. Then its tangent space at $\bar{x}$, $T_{\bar{x}} M$, is m-dimensional, and all the $\omega_{\bar{x}}^j$ must vanish on $T_{\bar{x}} M$, because of formula (25). Any subspace $E \subset T_{\bar{x}} M$ with this property will be called an *integral element* of system (24) at $\bar{x}$. The set of all m-dimensional integral elements at $\bar{x}$ will be

$$G_{\bar{x}}^m = \left\{ E \; \middle| \; \begin{array}{l} E \subset T_{\bar{x}} M \quad \text{and dim} \quad E \leq m \\ \omega_{\bar{x}}^1, \ldots, \omega_{\bar{x}}^K \quad \text{vanish on } E \end{array} \right\}.$$

Our first necessary condition is clear:

(26) $$G_{\bar{x}}^m \neq \varnothing.$$

7.3. – Differential ideals

To get the second one, let us ask a strange question: have we written all the equations? In other words, does the system

(27) $$\omega^1 = 0, \ldots, \omega^K = 0$$

exhibit all the relevant information?

The answer is negative. To see why, recall that M is a submanifold of $\mathbb{R}^n$, and denote by $i_M : M \to \mathbb{R}^n$ the standard embedding $i_M(x) = x$ for all $x \in M$. M is an integral manifold of system (27) if

(28) $$i_M^* \omega^1 = 0, \ldots, i_M^* \omega^K = 0.$$

But we know that exterior differentiation is natural with respect to pullbacks, that is, d commutes with i_M^*. So (28) implies

$$i_M^*(d\omega^1) = 0, \ldots, i_M^*(d\omega^K) = 0.$$

In other words, M is also an integral manifold of the larger system:

$$
\text{(29)} \qquad
\begin{cases}
\omega^1 = 0, \ldots, \ \omega^K = 0, \\
d\omega^1 = 0, \ldots, d\omega^K = 0,
\end{cases}
$$

which is different from (27). So integral elements of (29) will be different from integral elements of (27), and it is not clear which ones we should be working with.

To resolve this quandary, we shall assume that systems (27) and (29) have the same integral elements. In other words, the second equations in (29) must be *algebraic* consequences of the first ones. The precise statement for this is that the ω^k, $1 \leq k \leq K$, must generate a differential ideal:

$$
\text{(30)} \qquad \forall k, \ d\omega^k = \sum \alpha_j^k \wedge \omega^j .
$$

Note that if the given family $\{\omega^k \mid 1 \leq k \leq K\}$ does not satisfy this condition, the enlarged family $\{\omega^k , \ d\omega^k \mid 1 \leq k \leq K\}$ certainly will (because $dd\omega^k = 0$). So the condition that the ω^k must generate a differential ideal can be understood as saying that the enlargement procedure has already taken place.

Then our second necessary condition is that the ω^k, $1 \leq k \leq K$, must generate a differential ideal.

7.4. – A counterexample

It turns out that conditions (26) and (30) are almost sufficient. All we have to do is to replace (26) by a slightly stronger condition.

To see that (26) is not sufficient, let us give a simple example. Consider two functions f and g from $\mathbb{R}^{n-1}$ into itself, with $f(0) = g(0) = v \neq 0$. Define $\alpha^i(x, t)$ and $\beta^i(x, t)$, $1 \leq i \leq n - 1$, by

$$
\alpha^i(x, t) = f^i(x)dt - dx^i ,
$$

$$
\beta^i(x, t) = g^i(x)dt - dx^i ,
$$

and consider the exterior differential system in $\mathbb{R}^n$:

$$
\alpha^i(x, t) = 0, \quad \beta^i(x, t) = 0, \quad 1 \leq i \leq n - 1.
$$

The α^i and the β^i generate a differential ideal, and there is an integral element at 0, namely the line carried by $(1, N)$, so $G_0^1 \neq \emptyset$. However, finding an integral manifold of the initial system containing 0 amounts to finding a common solution of the two Cauchy problems:

$$(31) \qquad\qquad \frac{dx}{dt} = f(x), \quad x(0) = 0,$$

$$(32) \qquad\qquad \frac{dx}{dt} = g(x), \quad x(0) = 0,$$

which does not exist in general. So, though (26) is fulfilled, it is still impossible to integrate.

The counter example can also be restated in a more traditional way. Consider (31) and (32) as standard ODE, that have to be solved locally around 0. The natural first step is to linearize this system at 0 — which, incidentally, is exactly what the α^i and β^i do. Now, it turns out that the linearized version, at 0, does have a solution — namely

$$\frac{dx}{dt}(0) = v.$$

So we are in a case where the linearized problem has a solution, but the latter do not locally extend. Why is that? Clearly, the problem is that the equality $f(x) = g(x)$ — which is necessary for the existence of a solution — holds at $x = 0$, but not in the neighbourhood; the technical translation being that 0 is not an *ordinary point*. The interpretation, in this basic example, is quite simple. Consider the space $E(x, t)$ spanned by the forms $\alpha^i(x, t)$ and $\beta^i(x, t)$. At any point but $(0, 0)$, the dimension of $E(x, t)$ is 2. At $(0, 0)$, however, we have that $\dim E(0, 0) = 1$. In other words, at $(0, 0)$ we have a discontinuous drop in the dimension of the space. This is why $(0, 0)$ is not an ordinary point.

Another, less intuitive, counter-example goes as follows. Given a smooth function $f(x, y)$ of two variables, find two functions of one variable, $\phi(x)$ and $\psi(y)$, such that $f(x, y) = \phi(x) + \psi(y)$. This can be written as an exterior differential system in the submanifold M of $\mathbb{R}^4$ defined by the equation

$$(33) \qquad\qquad M = \{x, y, \phi, \psi \mid \phi + \psi = f(x, y)\}.$$

The corresponding exterior differential system is simply:

$$(34) \qquad\qquad d\phi \wedge dx = 0,$$

$$(35) \qquad\qquad d\psi \wedge dy = 0,$$

$$(36) \qquad\qquad dx \wedge dy \neq 0.$$

The reader will check easily that the system is closed, and that there is an integral element at every point. In addition, there is no clear singularity, as in the preceding example. And yet, if $\frac{\partial^2 f}{\partial x \partial y} \neq 0$ there is no integral manifold.

In general, what we need is a regularity condition which will exclude such pathological situations. Technically, this condition will state that the relevant equality holds true at *ordinary points*. Of course, this requires a general definition of the concept of an ordinary point. This is quite easy in the case of 1-forms, but may be more complex in general. This is the topic of the next subsection.

7.5. – Regularity conditions and ordinary points

If all the $\omega_k(x)$ are 1-forms, the regularity condition is clear enough: the dimension of the space spanned by the $\omega_k(x)$, $1 \leq k \leq K$, in $T_x^* \mathbb{R}^n$, should be constant on a neighborhood of $\bar{x}$ (which is obviously not the case in the counterexample above). Note that, locally, this dimension can only increase, that is, the codimension can only decrease.

If some of the ω_k have higher degree, the regularity condition is more complicated, because several codimensions are involved. However, the idea is the same: all these codimensions should be constant in a neighborhood of $\bar{x}$, which amounts to say that they should have the lowest possible value at $\bar{x}$. This is expressed in the following.

Pick a point $\bar{x} \in \mathbb{R}^n$; from now on, we work in the tangent space $V = T_{\bar{x}} \mathbb{R}^n$. Let $E \subset V$ be an m-dimensional integral element at $\bar{x}$. Pick a basis $\bar{\alpha}_1, \ldots, \bar{\alpha}_n$ of V^* such that:

$$E = \{\xi \in V \mid \langle \xi, \bar{\alpha}_i \rangle = 0 \quad \forall i \geq m + 1 \}.$$

For $n' \leq n$, denote by $\mathcal{I}(n', d)$ the set of all ordered subsets of $\{1, \ldots, n'\}$ with d elements. Denote by d_k the degree of ω_k. For every k, writing $\omega_k(\bar{x})$ in the $\bar{\alpha}_i$ basis, we get

$$\omega_k(\bar{x}) = \sum_{I \in \mathcal{I}(n, d_k)} c_I^k \, \bar{\alpha}_{i_1} \wedge \ldots \wedge \bar{\alpha}_{i_{d_k}}.$$

In this summation, it is understood that $I = \{i_1, \ldots, i_{d_k}\}$. Since $\omega^k(\bar{x})$ vanishes on E, each monomial must contain some $\bar{\alpha}_i$ with $i \geq$

$m + 1$. Let us single out the monomials containing one such term only. Regrouping and rewriting, we get the expression:

$$\omega^k(\bar{x}) = \sum_{J \in \mathcal{I}(m, d_k - 1)} \bar{\beta}^k_J \wedge \bar{\alpha}_{j_1} \wedge \ldots \wedge \bar{\alpha}_{j_{d_k - 1}} + \text{remainder},$$

where $\bar{\beta}^k_J$ is a linear combination of the $\bar{\alpha}_i$ for $i \geq m + 1$, and all the monomials in the remainder contain $\bar{\alpha}_i \wedge \bar{\alpha}_{i'}$ for some $i > i' \geq m + 1$.

Define an increasing sequence of linear subspace $H_0^* \subset H_1^* \subset \ldots \subset H_M^* \subset V^*$ as follows:

$$H_m^* = \text{Span}[\ \bar{\beta}^k_J \ | \ 1 \leq k \leq K, \ J \in \mathcal{I}(m, d_k - 1)\},$$
$$H_{m-1}^* = \text{Span}[\ \bar{\beta}^k_J \ | \ 1 \leq k \leq K, \ J \in \mathcal{I}(m - 1, d_k - 1)\},$$
$$H_0^* = \text{Span}[\ \bar{\beta}^k_J \ | \ 1 \leq k \leq K, \ J \in \mathcal{I}(0, d_k - 1)\}.$$

The latter is just the linear subspace generated by those of the $\omega^k(\bar{x})$ which happen to be 1-forms. We define an increasing sequence of integers $0 \leq c_0(\bar{x}, E) \leq \ldots \leq c_m(\bar{x}, E) \leq n$ by:

$$c_i(\bar{x}, E) = \dim H_i^*.$$

We are finally able to express Cartan's regularity condition. Denote by $\mathbb{P}^m(\mathbb{R}^n)$ the set of all m-dimensional subspaces of $\mathbb{R}^n$ with the standard (Grassmannian) topology: it is known to be a manifold of dimension $m(n - m)$. Denote by G^m the set of all (x, E) such that E is an m-dimensional integral element at x. Note that G^m is a subset of $\mathbb{R}^n \times \mathbb{P}^m(\mathbb{R}^n)$.

DEFINITION 7.2. Let $(\bar{x}, \bar{E}) \in G^m$ — that is, $\bar{E}$ is an m-dimensional integral element at $\bar{x}$. We say that $(\bar{x}, \bar{E})$ is *ordinary* if there is some neighborhood $\mathcal{U}$ of $(\bar{x}, \bar{E})$ in $\mathbb{R}^n \times \mathbb{P}^m(\mathbb{R}^n)$ such that $G^m \cap \mathcal{U}$ is a submanifold of $\mathbb{R}^n \times \mathbb{P}^m(\mathbb{R}^n)$ with codimension

$$c_0(\bar{x}, \bar{E}) + \ldots + c_{m-1}(\bar{x}, \bar{E}).$$

If all the ω^k are 1-forms, denote by $d(x)$ the dimension of the space spanned by the $\omega^k(x)$. Then $c_i(x, E) = d(x)$ for every i, and $(\bar{x}, \bar{E})$ is ordinary if $G^m \cap \mathcal{U}$ is a submanifold of codimension $md(\bar{x})$ in $\mathbb{R}^n \times \mathbb{P}^m(\mathbb{R}^n)$. This implies that, for every x in a neighborhood of $\bar{x}$, the set of $E \in G^m_x$ (integral elements at x) has codimension $md(\bar{x})$ in $\mathbb{P}^m(\mathbb{R}^n)$. It can be seen directly to have codimension $md(x)$. So

$d(x) = d(\bar{x})$ in a neighborhood of $\bar{x}$: this is exactly the regularity condition we wanted for 1-forms.

In the general case, if $(\bar{x}, \bar{E})$ is ordinary, the numbers c_i will also be locally constant:

$$c_i(x, E) = c_i(\bar{x}, \bar{E}) = c_i \quad \forall (x, E) \in \mathcal{U}.$$

The (non-negative) numbers

$$s_0 = c_0,$$
$$s_i = c_i - c_{i-1} \quad \text{for} \quad 1 \le i < m,$$
$$s_m = n - m - c_{n-1},$$

are called the *Cartan characters*. We shall use them later on.

7.6. – The main result

We are now in a position to state the Cartan-Kähler theorem:

THEOREM 7.1 (Cartan-Kähler). *Consider the exterior differential system*

$$\omega^k = 0, \quad 1 \le k \le K.$$

Assume that the ω^k are real analytic and that they generate a differential ideal. Let $\bar{x}$ be a point and let $\bar{E}$ be an m-dimensional integral element at $\bar{x}$. Assume $(\bar{x}, \bar{E})$ is ordinary. Then there is a real m-dimensional analytic integral manifold M, containing $\bar{x}$ and such that

$$T_{\bar{x}} M = \bar{E}.$$

This result is due to Elie Cartan in the case when the ω^k have degree 1 or 2, and was extended by Kähler to the general case; the reader may refer to [4] or [5] for a proof. Nothing should come as a surprise in this statement, except the real analyticity. It comes from the very generality of the Cartan-Kähler theorem. Indeed, every system of partial differential equations can be written as an exterior differential system.

To conclude, let us mention the question of uniqueness. There is no uniqueness in the Cartan-Kähler theorem: there may be infinitely many analytic integral manifolds going through $\bar{x}$ and having $\bar{E}$ as a tangent space at $\bar{x}$. However, the theorem describes in a precise way (not given here) the set

$$
\mathcal{M}_{\mathcal{U}} = \left\{ M \,\middle|\, \begin{array}{l} M \text{ is an integral manifold} \\ \text{and there exists } (x, E) \in \mathcal{U} \\ \text{such that } x \in M \text{ and } T_x M = E \end{array} \right\},
$$

where $\mathcal{U}$ is a suitably chosen neighborhood of $(\bar{x}, \bar{E})$. Loosely speaking, each M in $\mathcal{M}_{\mathcal{U}}$ is completely determined by the (arbitrary) choice of s_m analytic functions of m variables, the s_m being the Cartan character.

CHAPTER 8

Markets

In this concluding section we will give some indications about *markets*. It is assumed that there are n consumers and l goods, and all the goods are private. Each consumer has a certain amount of non-monetary resources $\omega_i \in \mathbb{R}^l_+$ and he has to solve the following problem:

$$(\mathcal{P}_{i,p}) \begin{cases} \max U_i(x_i), \\ p \cdot x_i \leq p \cdot \omega_i. \end{cases}$$

The quantity $x_i(p) - \omega_i$ is called *excess demand of the i-th agent* whereas $Z(p) := \sum_i x_i(p) - \sum_i \omega_i$ is called *excess demand*. Finally the quantity $\Omega := \sum_i \omega_i$ is the total quantity of goods in the market. If $Z \equiv 0$ then the market is said *at equilibrium*.

The problem we would like to solve is, as usual, the following:

Given a function $Z : \mathbb{R}^l \to \mathbb{R}^l$ is this an excess demand?

The answer to this problem was given by Sommerschein ([23], [24], [25]), Mantel ([17], [18], [19]) and Debreu ([11]) in a series of papers around 1974 and it is as follows: *if Z is a continuous function on $\mathbb{R}^l$, $n \geq l$, such that $p \cdot Z(p) \equiv 0$, then Z is an excess demand function. Moreover the utility functions U_i can be chosen to be concave.* Note that up to now we have not treated the question of concavity of utility functions, as it was required by the second question given in the first section. Another fact which it is important to say about the previous result is that the answer is not local, but global (in all $\mathbb{R}^l$).

However, another question remained open since that time, and it is the question of *markets* (not excess) *demand functions*. This means that each consumer has a fixed amount of money which he spends, and one observes the corresponding demand.

$$(\tilde{\mathcal{P}}_{i,p}) \begin{cases} \max U_i(x_i), \\[2mm] p \cdot x_i \leq 1. \end{cases}$$

The optimal consumptions $x_i(p)$ are called *individual functions demands* and summing up we get the collective market demand function $X(p) = \sum_i x_i(p)$. We will prove now the following

THEOREM 8.1 (Chiappori-Ekeland). *Let* $\bar{p} \in \mathbb{R}^l$, $n \geq 1$, $\mathcal{U}$ *a neighbourhood of* $\bar{p}$, $X : \mathcal{U} \to \mathbb{R}^l$ *real analytic such that*

$$p \cdot X(p) = 1, \quad \forall p \in \mathcal{U}.$$

Then there exists utility functions $U_i : \mathbb{R}^l \to \mathbb{R}$, $\hat{x}_i : \mathcal{U} \to \mathbb{R}^l$, *real analytic for* $i = 1, \ldots, n$, *such that:*

(i) $X(p) = \sum_i \hat{x}_i(p)$;
(ii) $\hat{x}_i(p)$ *solves* $(\tilde{\mathcal{P}}_{i,p})$;
(iii) *the* U_i *are concave in a neighbourhood of* $\hat{x}_i(p)$.

PROOF. We start by translating the conditions (i), (ii), (iii) in the language of the differential forms. We will end up with an exterior differential system.

Let us introduce the indirect utilities functions:

$$V_i(p) = U_i(\hat{x}_i(p)) = \max \{U_i(x_i) + \lambda_i(1 - p \cdot x_i)\}.$$

Then

$$\nabla V_i(p) = \lambda_i(p)\hat{x}_i(p),$$

and in terms of the known function X we have the equation

$$(37) \qquad X(p) = -\sum_{i=1}^{n} \frac{1}{\lambda_i(p)} \nabla V_i(p), \quad p \in \mathbb{R}^l,$$

or, in other words, once again the problem to decompose a given function as a linear combination of n gradients. Moreover, by the constraint $p \cdot \hat{x}_i(p) = 1$ there will be the equations

$$(38) \qquad p \cdot \nabla V_i(p) = -\lambda_i(p), \quad i = 1, \ldots, n.$$

Finally the third conditions of Theorem 8.1 is satisfied if the matrix of the second derivatives $\nabla \nabla V_i(p)$ is positive definite.

Now we introduce the manifold M in $\mathbb{R}^l \times \mathbb{R}^{ln} \times \mathbb{R}^n$ defined by:

$$M := \left\{ (x^i, p_i^k, \lambda_k) : x_i = -\sum_{k=1}^{n} \frac{1}{\lambda_k} p_i^k, \quad \sum_{i=1}^{l} x_i p_i^k = -\lambda_k, \quad \forall i, k \right\}.$$

The meaning of the manifold M is clear. Note that M is defined by $n + l$ algebraic equation. Now we would like to have:

(1) p_i^k as functions of $(x_1, \dots, x_n)$;

(2) concave functions V_k such that $p_i^k = \dfrac{\partial V_k}{\partial x_i}$.

In the language of differential forms (2) is equivalent to

$$dV_k = \sum_{i=1}^{l} p_i^k dx_i,$$

and by the Poincaré lemma

$$\sum_{i=1}^{l} dp_i^k \wedge dx_i = 0.$$

This condition is equivalent to the following problem: find an n-dimensional integral manifold for the system

(39) $$\sum_{i=1}^{l} dp_i^k \wedge dx_i = 0, \quad k = 1, \dots, n.$$

This problem is solved by applying the Cartan-Kähler theorem. One must then show that the function V_i we obtain can be taken to be concave. We refer to [7] for the full proof.   □

References

[1] G. B. ANTONELLI, *Sulla Teoria Matematica della Economia Politica*, 31 pp. Pisa (Tipografia del Folchetto), 1886 (reprinted in facsimili in Giornale degli Economisti e Statistica Matematica **10** (1951), 233-263).

[2] V. ARNOLD, "Mathematical Methods of Classical Mechanics", Naouk, Moscow, 1965 French translation, Editions de Moscou, 1968; English translation, Springer, 1975.

[3] M. BROWNING —P. A. CHIAPPORI, *Efficient intra-household allocations: a general characterization and empirical tests*, Econometrica **66** (1998), 1241-1278.

[4] R. BRYANT —S. CHERN —R. GARDNER —H. GOLDSCHMIDT —P. GRIFFITHS, "Exterior Differential Systems", MSRI Publications vol. 18, Springer-Verlag, 1991.

[5] E. CARTAN, "Les systémes différentiels extérieurs et leurs applications géométriques", Hermann, 1945.

[6] P. A. CHIAPPORI —I. EKELAND, "Problémes d'agrégation en théorie du consommateur et calcul différentiel extérieur", CRAS Paris (323), pp. 565-570, 1996.

[7] P. A. CHIAPPORI —I. EKELAND, *Aggregation and market demand: an exterior differential calculus viewpoint*, Econometrica **67** (1999), 1435-1458.

[8] P. A. CHIAPPORI —I. EKELAND, *Disaggregation of collective excess demand functions in incomplete markets*, preprint, CEREMADE and DELTA.

56

[9]   P. A. Chiappori —I. Ekeland —F. Kubler —H. Polemarchakis, *The identification of preferences from equilibrium prices*, submitted.

[10] P. A. Chiappori —I. Ekeland, *A convex Darboux theorem*, Ann. Scuola Norm. Sup. Pisa Cl. Sci. **25** (1997), 287-297.

[11] G. Debreu, *Excess demand functions*, Journal of Mathematical Economics **1** (1974), 15-23.

[12] W. E. Diewert, *Generalized Slutsky conditions for aggregate consumer demand functions*, Journal of Economic Theory **15** (1977), 353-362.

[13] J. Genakoplos —H. Polemarchakis, *On the disaggregation of excess demand functions*, Econometrica, (1980), 315-331.

[14] P. Griffiths —G. Jensen, *Differential systems and isometric embeddings*, Princeton University Press, 1987.

[15] D. M. Kreps, "Microeconomic Theory", Longman, 1993.

[16] D. McFadden —A. Mas-Collel —R. Mantel —M. Richter, *A characterization of community excess demand fuctions*, Journal of Economic Theory **7** (1974), 361-374.

[17] R. Mantel, *On the characterization of aggregate excess demand*, Journal of Economic Theory **7** (1974), 348-353.

[18] R. Mantel, *Homothetic preferences and community excess demand functions*, Journal of Economic Theory 12 (1976), 197-201.

[19] R. Mantel, *Implications of economic theory for community excess demand functions*, Cowles Foundation Discussion Paper n. 451, Yale University, 1977.

[20] H. M. Polemarchakis, *Disaggregation of excess demand under additive separability*, European Economic Review **20** (1983), 311-318.

[21] E. Slutski, *Sulla teoria del bilancio del consumatore*, Giornale degli Economisti e Rivista di Statistica (3) **51** (1915), 1-26.

[22] W. Shafer —H. Sonnenschein, *Market demand and excess demand functions*, chapter 14 In: "Handbook of Mathematicla Economics", K. Arrow and M. Intriligator (eds), vol. 2, North Holland, Amsterdam, 1982, pp. 670-693.

[23] H. Sonnenschein, *Market excess demand functions*, Econometrica **40** (1972), 549-563.

[24] H. Sonnenschein, *Do Walras'identity and continuity characterize the class of community excess demand functions?*, Journal of Economic Theory 6 (1973), 345-354.

[25] H. Sonnenschein, *The utility hypothesis and market demand theory*, Western Economic Journal **11** (1973), 404-410.

PUBBLICAZIONI DELLA CLASSE DI SCIENZE
DELLA SCUOLA NORMALE SUPERIORE
QUADERNI

1. DE GIORGI E., COLOMBINI F., PICCININI L.C.: *Frontiere orientate di misura minima e questioni collegate.*
2. MIRANDA C.: *Su alcuni problemi di geometria differenziale in grande per gli ovaloidi.*
3. PRODI G., AMBROSETTI A.: *Analisi non lineare.*
4. MIRANDA C.: *Problemi in analisi funzionale* (ristampa).
5. TODOROV I.T., MINTCHEV M., PETKOVA V.B.: *Conformal Invariance in Quantum Field Theory.*
6. ANDREOTTI A., NACINOVICH M.: *Analytic Convexity and the Principle of Phragmén-Lindelöf.*
7. CAMPANATO S.: *Sistemi ellittici in forma divergenza. Regolarità all'interno.*
8. TOPICS IN FUNCTIONAL ANALYSIS: *Contributors:* F. STROCCHI, E. ZARANTONELLO, E. DE GIORGI, G. DAL MASO, L. MODICA.
9. LETTA G.: *Martingales et intégration stochastique.*
10. OLD AND NEW PROBLEMS IN FUNDAMENTAL PHYSICS: *Meeting in honour of* G.C. WICK.
11. INTERACTION OF RADIATION WITH MATTER: *A Volume in honour of* ADRIANO GOZZINI.
12. MÉTIVIER M.: *Stochastic Partial Differential Equations in Infinite Dimensional Spaces.*
13. SYMMETRY IN NATURE: *A Volume in honour of* LUIGI A. RADICATI DI BROZOLO.
14. NONLINEAR ANALYSIS: *A Tribute in honour of* GIOVANNI PRODI.
15. LAURENT-THIÉBAUT C., LEITERER J.: *Andreotti-Grauert Theory on Real Hypersurfaces.*
16. ZABCZYK J.: *Chance and Decision. Stochastic Control in Discrete Time.*

CATTEDRA GALILEIANA

1. LIONS P.L.: *On Euler Equations and Statistical Physics.*

LEZIONI LAGRANGE

1. VOISIN C.: *Variations of Hodge Structure of Calabi-Yau Threefolds.*

LEZIONI FERMIANE

1. THOM R.: *Modèles mathématiques de la morphogénèse.*
2. AGMON S.: *Spectral Properties of Schrödinger Operators and Scattering Theory.*
3. ATIYAH M.F.: *Geometry of Yang-Mills Fields.*
4. KAC M.: *Integration in Function Spaces and Some of Its Applications.*
5. MOSER J.: *Integrable Hamiltonian Systems and Spectral Theory.*
6. KATO T.: *Abstract Differential Equations and Nonlinear Mixed Problems.*
7. FLEMING W.H.: *Controlled Markov Processes and Viscosity Solution of Nonlinear Evolution Equations.*
8. ARNOLD V.I.: *The Theory of Singularities and Its Applications.*
9. OSTRIKER J.P.: *Development of Larger-Scale Structure in the Universe.*
10. NOVIKOV S.P.: *Solitons and Geometry.*
11. CAFFARELLI L.A.: *The Obstacle Problem.*

ALTRE PUBBLICAZIONI

Proceedings of the Symposium on FRONTIER PROBLEMS IN HIGH ENERGY PHYSICS
Pisa, June 1976

Proceedings of International Conferences on SEVERAL COMPLEX VARIABLES, Cortona,
June 1976 and July 1977

*Raccolta degli scritti dedicati a JEAN LERAY apparsi sugli Annali della Scuola Normale
Superiore di Pisa*

*Raccolta degli scritti dedicati a HANS LEWY apparsi sugli Annali della Scuola Normale
Superiore di Pisa*

*Indice degli articoli apparsi nelle Serie I, II e III degli Annali della Scuola Normale
Superiore di Pisa* (dal 1871 al 1973)

*Indice degli articoli apparsi nella Serie IV degli Annali della Scuola Normale Superiore
di Pisa* (dal 1974 al 1990)

ANDREOTTI A.: *SELECTA vol. I, Geometria algebrica.*

ANDREOTTI A.: *SELECTA vol. II, Analisi complessa, Tomo I e II.*

ANDREOTTI A.: *SELECTA vol. III, Complessi di operatori differenziali.*

Fotocomposizione "CompoMat" Loc. Braccone, 02040 Configni (RI), Italy
Finito di stampare per conto della "CompoMat" dalla Nuova Grafica 86 nel gennaio 2001